MW01103321

Wireless Data for the Enterprise
Making Sense of Wireless Business

Authors:

George S Faigen
Chief Marketing and Strategy Officer,
Broadbeam Corporation

Boris Fridman
Chairman and CEO,
Broadbeam Corporation

Editor:
Arielle Emmett

TO:

We dedicate this book to our loving families who were so supportive during this effort.

George S Faigen:
To my parents, Sonny and Bernice
To my wife Naomi and
To my children Jordan and Avery
You inspire and guide everyday of my life.

Boris Fridman:
With love and affection, to my mother Khana Fridman,
my wife Ruth, and my two sons Mark and Ben.

Arielle Emmett:
For my children, Grainne and Emmett Arthur, with love forever.

McGraw-Hill

A Division of The **McGraw·Hill** *Companies*

1 2 3 4 5 6 7 8 9 0 DOC/DOC 0 9 8 7 6 5 4 3 2 1

ISBN 0-07-138637-8

The sponsoring editor of this book was Judy Bass, and the production supervisor was Sherri Souffrance.
This book was set in Rotis and Univers.
Printed and bound by R. R. Donnelley & Sons Company.

This book is printed on recycled, acid-free paper containing a minimum of 50% recycled, acid-free stock.

Book Design: Circle Square and Triangle Graphic Designs, Inc.
Creative Director: Susan Cotler-Block
Designers: Lillian Ng, Rosemary Markowsky, Preeti Monga

CONTENTS

FOREWORD

The proliferation of wireless data across vertical and horizontal businesses, as well as with consumers, marks a profound change in the way organizations conduct business and people engage in their everyday activities. Businesses today are just beginning to recognize the importance of mobile workers and the impact these employees have on the enterprise. With the introduction of laptops into the workforce as replacements for PCs, the potential of this new type of worker became evident. Although it is taking many years and the journey is ongoing, these workers are finally achieving the first-class status they warrant as business employees. Now, given the growing number of mobile workers in the United States, Europe, Asia, and Latin America; the expansion in the number and types of mobile devices; and the abundance of software and services available to deploy wireless data networks, the once meek and under-appreciated mobile worker shall indeed inherit the corporate earth.

Vertical industries have been part of the wireless landscape for many years because of a visible need to provide productive workers with a means to perform their daily tasks more efficiently. For example, in order to realize increased productivity and improved customer service, package delivery companies, utilities, and service-oriented organizations sought out and deployed wireless solutions as early as 15 years ago.

The road for horizontal enterprise mobilization projects, however, has just begun to be paved. In the 1990s, when the Internet was introduced to businesses to play an important role in extending existing corporate resources to employees, partners, and customers, much of the focus was on an improved Web and e-commerce presence. Just as companies sought to leverage the Internet to access their corporate intranets then, businesses today are presented with the opportunity to extend the reach of their critical data to the wireless world.

The opportunity to travel this road is one filled with many promises of significant return on investment (ROI) through enhanced productivity, reduction of costs, improved customer service, and increased revenue. But the critical decision to wirelessly enable applications for employees, partners, or customers is a difficult one that requires effective project planning. The way this decision is made in an organization could mean the difference between a successful, scalable, secure, and usable solution and a single-minded one that has limited functionality and is under-utilized. Typically, many of the decisions to "go wireless" come down from a CEO or other business line manager. The thought is that something must be done regarding wireless, but the exact project may be a challenge to identify if it is not clearly defined. Justification for the business case and the technical requirements of wireless deployment involve multiple factors that can affect the entire organization for many years to come.

As wireless networks evolve and new devices and software solutions enter the market, much education is required on the part of the enterprise to determine how it will "go wireless." Organizations must begin the decision-making process by building the case for deploying wireless solutions and then tackle the issues of identifying the target recipients of data, determining a method of deployment, setting a time for the rollout, and choosing suppliers. Complete guidance in this decision-making process is sometimes absent from vendors selling solutions that address one particular area of wireless deployment. Encompassing these requirements and taking a holistic view beyond single-function deployments as a way to broaden the sometimes nearsightedness of the mobilization project scope are fundamental in choosing the right supplier(s) and deploying successful wireless solutions.

Enterprise wireless deployments are still in their early stages as wireless networks evolve, numerous mobile device types flood the market, and businesses become educated about the value of wireless middleware and the importance of surrounding partners, including system integrators, professional services organizations, and application vendors. Solutions available today offer mobilization projects that bring the reality of wireless enablement to the forefront, albeit with limits. Enhancements across the entire spectrum of solutions allow organizations to continue their journey, stirring up creative juices, developing more robust applications, and expanding opportunities for new wireless deployments not yet dreamed of.

Wireless Data For The Enterprise: Making Sense of Wireless Business addresses the most important questions that all organizations should ask themselves when they consider deploying wireless applications. It looks into a not-too-distant future, with a description of a ubiquitous wireless world. The authors then provide meticulous details regarding existing wireless use, highlight important historical and technical background, and deliver a lively set of wireless deployment case studies. The authors also convey the essential discussion surrounding technology, including middleware, mobile devices, and networks, and they offer a business recipe for cooking up a wireless project plan.

This book is a valuable tool for management considering or already involved in wireless project planning as well as an important guide for IT personnel charged with leading their company through wireless deployments. Others involved in the financial, technology, and partner channel communities will also find it helpful to sift through much of the hype surrounding the wireless universe.

Stephen D. Drake, Research Manager
Wireless and Mobile Enterprise Access
IDC

ACKNOWLEDGMENTS

This book could not have been written without the involvement and assistance from many individuals, from Broadbeam employees who contributed to the knowledge and development of the material, to customers and wireless innovators who were willing to share their stories, to wireless industry pundits and experts whose research and opinions have impacted the authors along this exciting journey. We would like to thank the following individuals and companies for their involvement in this groundbreaking effort:

Federal Express Corporation, especially Winn Stephenson, Senior Vice President of Information Technology; Fidelity Investments Institutional Services Company, especially Joseph Ferra, Senior Vice President and Chief Wireless Officer; BellSouth Corporation, especially Gary Dennis, General Manager, Telecommunications Network Operations Group; Sears, Roebuck and Company; Northeast Utilities, especially Andy Kasznay, Software Engineer; Aeris.net; Notifact Corporation, especially David Sandelman, Vice President and Chief Technical Officer; London (Ontario) Police Service, especially Eldon Amoroso, Director of Information and Technology; Versaterm, Inc.; ADP Claims Solutions Group; COUNTRY Insurance and Financial Services; Orange Intelligent Home Research Centre; Symbol Technologies Inc.; and Itronix Corporation.

Also, Stephen D. Drake, Research Manager, Wireless and Mobile Enterprise Access, IDC, Carolyn Joiner, Broadbeam and Susan-Cotler Block, Circle Square and Triangle, for her creative book design. Finally, Tamara Gruber, Broadbeam, whose project management, tireless effort, and sound advice made this book a reality.

Photo Credits: Patrice Elmi, Japan National Tourist Organization, Rosemary Markowsky, NASA, Lillian Ng.

INTRODUCTION

We set out to write this book in order to deliver a definitive blueprint for thinking about and implementing technologies required to "go wireless," helping enterprises extend their knowledge flows to encompass mobile workers, customers, and suppliers, and to develop new models for profit and selective data exchange.

We believe that this book will help corporate managers develop a sense of what is possible with wireless today – to tailor personalized data for users while selecting wireless devices, gateways and networks. It will show the technical reader how to implement wireless applications that enable disruptive market strategies and new methods for generating revenue, linking enterprise assets securely with people using laptops, personal digital assistants (PDAs) smart phones, palmtops, dashboards, and other varying flavors of mobile devices.

Most importantly, the book provides a reality check for controversial claims about wireless. For the management reader, it uncovers what can be accomplished effectively today in both the business and consumer space using existing applications and networks and outlines the most immediate possibilities for a fully untethered enterprise and lifestyle tomorrow. It guides a reader on a journey beyond the Internet and simple cellular systems to a world of wireless knowledge driving the enterprise to new levels of performance, mobile e-commerce, and customer satisfaction.

Wireless Data For The Enterprise: Making Sense of Wireless Business starts out in Chapter One, "A Day in the Life of Mobile She Devil," by describing a futuristic view of how wireless data will touch every aspect of our lives from home monitoring, to wireless coupons, to email access, to news alerts, to wireless business applications. The chapter concludes by looking at the trends effecting the implementation and effectiveness of wireless applications. To provide a backdrop for our discussion, Chapter Two "Needed and Do-Able: A Wireless Data History," walks readers through the early development of wireless data communications, from ancient smoke signals to today's wireless data networks, showing the acceleration of the development cycle and how needs drive technology. Then we explore how new drivers are pushing the growth of wireless business applications.

Chapter Three, "Untethered Enterprise: No Pain, No Gain," outlines lessons learned by early enterprise adopters and outlines questions to ask yourself to determine if a wireless application is right for your business. It provides three "Dead

Giveaways" for a wireless solution, helping managers and technical staff through the complex decision process that is always the beginning of a wireless implementation. "At the Leading Edge," Chapter Four, walks readers through a variety of enterprise wireless data case studies, included to show the many ways wireless can be used to improve business processes and impact the bottom line. These case studies provide detail descriptions of the application development and deployment cycles, including advice on what they would do differently and what went right. The next chapter, "The Wireless Consumer: What Will It Take to Woo the Mass Market?," moves on to talk about the possibilities for pervasive computing and how wireless will change the way we work, play and live.

These introductory chapters set the stage for the meat of the book, which lies in Chapters Six through Nine. With Chapter Six, "Middleware is the Hub of Wireless Computing," we take a close look at what ties all the pieces of a solution together – wireless middleware. Wireless middleware is often overlooked when first designing a wireless application, but it is the crucial element that people come back to time and again to make their solution work successfully. Moving on, in Chapter Seven, "Wireless Networks and Devices Complete the Mobile Landscape," we begin to work through the elements of a wireless deployment, guiding business managers, as well as those charged with the development of the wireless application, through the alphabet soup that represents the wireless networks and devices. This is geared to help the reader understand the differences between network and device types and to determine which best fit their business need. Chapter Eight, "Creating a Wireless Business: A Primer," was written to walk the reader through each stage of a wireless deployment, making sure critical steps aren't overlooked, and understanding what questions to ask of vendors and what features to look for depending on the business need and user.

Finally, the book concludes with Chapter Nine, "Boldly Go...Inventing the Wireless Enterprise," which helps readers take all the knowledge they have gained thus far and decide how, where, and when to utilize wireless to improve their business and achieve a competitive advantage.

Wireless Data For The Enterprise: Making Sense of Wireless Business takes the unique position that untethering the enterprise holds the key to business process innovation, as well as new levels of wireless productivity and profitability – not a ness develops the following themes:

▶▶ By rethinking the business process(es) of the enterprise, IT managers can develop an effective plan for using wireless data today.

▶▶ Businesses need wireless to increase productivity, customer satisfaction, and competitive edge. Productivity means time savings, increased revenues, and happier customers: more service calls made, faster order processing, instantaneous package tracking, image transfer, dynamic dispatch, secure access to corporate legacy systems, and 'anytime, anywhere' e-commerce transactions.

▶▶ Wireless initially means higher pain and cost for early adopters, but technology responds rapidly to need. The pioneering vertical applications of wireless data are primary examples. Initially the services were limited in scope and costly. But with every iteration of hardware, middleware and apps, wireless-driven technology shows its potential for enhancing incumbent technology, even toppling it, with better technology.

▶▶ Wireless data creates 'disruptive' business models and new markets. It reengineers a process, improves one, creates new models for production, services and revenue generation, offering business and consumers something smaller, cheaper, better, and faster – ultimately disruptive to other business models and production flows. (See our Case Studies for examples.)

▶▶ Many business processes are dramatically improved by the acquisition and transmission of mobile data in real-time. Wireless data integrates both data collection and push services, an inflowing and outflowing of vital information that has bottom-line impact. An example: Auto accident insurance adjusters can use wireless terminals and claims processing on the scene to evaluate a claim, acquire centralized actuarial data and records, and also transmit and receive claims, cutting customers a check on site. The competitive advantages of an organizational process like this are enormous; customers flock to business models that serve them better and more directly.

The book outlines the history of wireless data and shows exactly how enterprise managers can develop effective models – and pilots – for mobility. The book portrays the road warrior, the connected enterprise, the challenge of accessing the wireless Internet and corporate Intranet – and covers wireless basics, standards issues, caveats and components of an effective mobile solution.

Most importantly, the book provides a vision of how to imagine and implement wireless requirements in places where no mobility solution has gone before. Aimed principally at the business thinker and corporate IT manager, it also contains advanced tools and tables for wireless applications developers and consultants.

ABOUT THE AUTHORS

Boris Fridman, Chairman and Chief Executive Officer of Broadbeam Corporation is a founder of Broadbeam Corp. He led the company to become a premier provider of software products designed to provide developers with tools to speed and ease the integration of wireless connectivity and solve common mobility challenges.

A wireless data industry visionary, Fridman is a frequent presenter at conferences including CTIA's Wireless show, DCI's Wireless Summit, Wireless IT, COMDEX, PC Expo, eCRM, Mobile Software Forum and Marcus Evans conferences. Fridman's bylined articles have appeared in publications including WirelessNow (online), Wireless Week, Frontline Solutions, Wireless Integration, Data Communications and others. He holds B.S. and M.S. degrees in physics from the University of Belorussia, Minsk, Belarus.

George S Faigen is Broadbeam's Chief Marketing & Strategy Officer. As CMSO, he drives the company's strategy as it rapidly expands its presence in new markets. As a veteran of 22 years in the computer industry, Faigen's skills and experience complement the talents of Broadbeam's senior management team.

Prior to joining Broadbeam, Faigen was the head of RONIN Consulting where he developed leading eMarketing strategies for computer and telecommunication firms. His pioneering work aided these firms in being able to identify and capture new markets for a wide range of hardware, software and services. Faigen also built the business case, positioning and launch plan for Web ASP 1by1.com. A frequent speaker at industry events, Faigen has presented at conferences including COMDEX Chicago, Marcus Evans conferences, the Wireless Symposium, Supercomm and Mobile Application Development.

Faigen previously worked as VP of Marketing at Wang Laboratories and at IBM, where he held positions in marketing, sales, product development and product research. He has an M.S. in mechanical engineering from the University of Kentucky, a B.S. in mechanical engineering from Carnegie-Mellon University and holds a Professional Engineering License.

Arielle Emmett, a former editor in chief of America's Network and Wireless Integration magazine, is a Sigma Delta Chi Award Winner for investigative journalism. She is the author of *Going Mobile: The Wireless Internet & Other Data Solutions* (CTIA/Wireless Data Forum, January 2001), and has contributed to The Los Angeles Times, Saturday Review, Ms., OMNI, The Village Voice, The Boston Globe, Parents Magazine, Data Communications, and Network World, among many others. Arielle has served as Taiwan correspondent for Newsweek and a lifestyle reporter for The Detroit Free Press. She lives with her children in Wallingford, Pennsylvania.

青龍
ワールド

1 A DAY IN THE LIFE OF A MOBILE SHE DEVIL

SHE WAKES AT 5AM IN A TOKYO HOTEL, jarred to alertness by the digital alarm. Without rising from the bed she is already hearing Tom Brokaw and Bill Gates chatting on a streaming video link from MSNBC directly to her palm-size *Personal Digital Assistant* (PDA), exhorting her that Wall Street is unhappy. The Federal Reserve Bank has not lowered rates; the stock market tanks; there is a crisis in the Middle East.

Her PDA is now synchronizing its *Personal Information Management* (PIM) software – schedules, sales contacts, addresses – wirelessly with her desktop computer in Los Angeles. Down comes her updated calendar, down comes a 'skinnied down' contact list she'll need today from her customer list in her *Customer Relationship Management* (CRM) system. More news flashes: *WORLD ECONOMY IS SHAKEN BY SLOW DOWN IN ASIA PACIFIC* plus half a dozen e-mails, a logistics report from a shipper that manages load tracking via the wireless web. *LOST PACKAGE*: her display reads. *DO YOU WANT THE TRACKING NUMBER?*

She rises now and drops her robe on the floor, enjoying an off-color note from her boyfriend appended to a Lotus Notes document (he works in the Hong Kong office). She is singing, "Shall We Dance?" showering to the tunes from NAPSTER downloaded to her PDA. Her world phone rings and synchronizes through a *Bluetooth* wireless connection with her laptop. The laptop is open, and she "reads" her morning mail. Another CNN news flash: *NASDAQ TAKES BEATING*; another four voice mails telling her the home office can't find that Osaka shipment. And now the hotel breakfast menu flashes on her PDA in English and Japanese, with prices marked clearly in Yen. Down comes the menu; down comes her link to a digital expense report; down comes her recorded preferences; will she stick with the same breakfast? Of course, with one exception (bacon). The FedEx tracking number of that lost package is "pushed" to her screen – where has the package gone? Through the surging droplets of water, she hears a growl: "Pick up the phone, damnit." It's her boss. She doesn't answer. She hears and sees his awful Max Headroom jabber on her PDA in another short video link as he exhorts the Tokyo sales team to make their numbers this month. She wonders if anyone has ever had cleaner fingernails. Another love note: Donald misses her so much – will she call tonight to help with homework? It's her ten year-old son.

Nothing is too far fetched for her. Mentally, she is hard wired to her PDA, her laptop, and her networks; everything else is wireless and

free. Here she can be a she devil, a road warrior, writing salacious love notes in waves of digital traffic, connecting and hopping on and off a taxi ride in an ethereal world as choked with messages as the Ginza at rush hour. Her words break up into garbled codes; her wireless terminal snips off pieces of conversation and data, sending them to mobile receiving stations and wireless base stations, then a Trans-Pacific fiber optic cable network, a satellite network, and finally wireless base stations again in the Los Angeles basin. Somewhat miraculously the networks reassemble her messages into coherent form. The PDA beeps again: hotel breakfast is ready; she dictates a note to her U.S. doctor, speaking directly into her PDA, which translates her voice into a text message, saying she is out of allergy medication, can he wire a secured prescription through customs to a local Tokyo pharmacy? She looks up a pharmacy in a directory of approved English-speaking retail services, then sends it along. Another link: Louis Rukeyser reports the NASDAQ will recover; tech investors have too much to lose. With a single push button she executes a trade to buy more Oracle and Nortel at bargain-basement prices. A local Nippon News Service "pushes" Osaka weather to her terminal; she will be headed there in three hours on the bullet train. Chilly, overcast skies, hotel reservations – she gets another room at the Osaka Hilton, debits her American Express Account, downloading the item to her expense report on the PDA.

Then she hits the road. She figures the cabby won't speak English if she burdens him with a confusing address, so she loads up her PDA with a Japanese-English mini dictionary module, looking up the *Global Positioning Satellite* (GPS) location for her first appointment. The driver chuckles as she hands him her display showing the directions in *kanji*. Another message from FedEx: it's the shipment of those *Third Generation* (3G) chipsets moving from a San Diego cargo bay to a 747 bound for Haneda, not Osaka. Her handheld drones: "You have Confirmation," flashing a digital signature showing proof of package delivery. Relief: her Japanese assistant can pick it up; she whips out her i-Mode phone and pages her. Now an e-coupon crops up on her terminal, offering her an extra 10 Yen off a Sushi McMuffin at McDonalds. She orders her uniquely Japanese snack by tapping her PDA with a stylus.

How could she live without all of this? How could she face the day? She cannot exist without wireless telemetry, instant messaging, connected organizers, streaming multimedia, access to corporate data behind the firewall, e-mails, www.travelocity.com, the Osaka weather report, the GPS directions to Kamakura, her wirelessly linked American Express account. Each day she carries a minimum of three, sometimes four wireless devices – interactive messenger, smart phone, PDA, laptop. Snatching her Gucci bag, she picks up her 2.2 pound laptop, wirelessly enabled with her tiny PC card modem; her terminals are snug in a cradled belt against her body: something like a holster. She steps out of the cab and e-mails her broker in Hong Kong to reserve that prime apartment she'll share with her son once she completes a corporate transfer in two months.

Half the world away, her son is shutting down the condo. His smart phone sensor is linked to a *Radio Frequency* (RF) residential gateway inside the garage; it monitors lights, electricity, water consumption, Internet access, voice and video communications, both wireless and wireline. Floodlights go on, interior lights doused, door locks secured; computer off; air conditioning set to a higher temperature automatically to accommodate the breezes from the San Gabriel Mountains. Energy read-outs are sent automatically to Mom in Tokyo. She can hear the barking of the two dogs as they wait patiently inside the garage for Donald to return. She checks the status of the security system at her home, gets a peek of the dogs captured on a digital camera, and hands the cabby her PDA once again – he guffaws as the dogs bark. Quick stop to pick up her Sushi McMuffin; the cabby gets a remote sensing message from his gas tank to pull over and fill up. As he heads for the gas station, careening around a corner, he nearly knocks over a kimono-clad Japanese lady doing her morning shopping. When he comes to an abrupt stop half a block away, a foreigner tailgating a little too close suddenly slams on his brake and rear-ends him. The two cars stop. She swears under her breath: *late*. Cabbie whips out a cell phone, calling NTT emergency services and insurance companies, inspecting the damage – just a few scrapes. Neither cabbie nor foreigners look at each other.

Within minutes a Japanese adjuster shows up, evaluates the cabbie's claim on the spot. A wireless terminal pops out an estimate and damage report, takes a digital picture, then sends the information directly to the insurance company's claims processing system. The estimator cuts an e-check on the spot, bows to everybody; the money is automatically deposited to the taxicab corporate account. Meanwhile, the foreigner is still waiting for a tow truck. He is frowning; no problem; wireless dispatch works beautifully, only Tokyo traffic is awful. Eventually he'll get help. She hops another cab, watching the sunrise. Her PDA beeps against her belt – another message – she cannot imagine a world any other way.

"Disruptive Technology" — A New Alchemy

It was, of course; and it is a world away. But while the world is not quite as freely connected as this imaginative scenario might suggest, the journey toward wireless data is proceeding with incredible speed. It is quite likely this Japanese scenario will seem too conservative – not outlandish enough – in years to come, as wireless data becomes an unbelievably pervasive influence across the planet.

Fifty years ago, Marshall McLuhan reinvented our thinking about connectedness when he wrote the book, *The Medium is the Message*. At the time, McLuhan alluded to the vast disruptions in cultural life made possible through the power of media and its chaotic exposures – mostly television, movies, and radio. Since then, we've lived beyond TV to see the death of the typewriter, the rise of the minicomputer, the advent of the standalone and networked PCs; the invention of the electronic spreadsheet, the database, the word processing program; the 56 *kilobyte per second* (Kbps) modem, the creation of the cordless phone, the ubiquity of the Internet, the rise of cable modems, *Digital Subscriber Line* (DSL), fiber, and satellite networks; the penetration of wireless *Local Area Networks* (LANs); and now, the launching of Web-connected wireless phones, laptops, and PDAs. This is only the beginning of a new set of disruptions that will transform business, markets, and the way we conduct our lives.

Disruption is an interesting word. In the language of Harvard Business School professor Clayton M. Christensen *(The Innovator's Dilemma, Harvard Business School Press, 1997)*, disruptive technologies represent innovations that appeal to emerging markets – not existing ones. They reinvent a whole new standard for how products and services operate, perform, and how customers use them, reshaping markets by changing equations for productivity, even life style.

Disruptive technologies, however, start out as innovations that result in *worse* product performance than existing sustaining technologies. Classically, disruptive technologies under-perform established products in mainstream markets. But they have features – think of the earliest wireless handsets and web phones – that a few 'fringe' customers value highly. Over time, products based on disruptive technologies reshape thinking; they cater to hidden needs; they suggest new ways of acting and transacting. Typically these products are cheaper, simpler, smaller, and frequently, more convenient to use than existing technologies. So, for example, just as the PC supplanted the minicomputer, so the laptop created a fresh market for computing on the go. Wireless in turn created mobile talking, then mobile data. And today, thanks to more powerful microprocessors and the advent of better networks with more flexible bandwidth allocation, wireless offers the promise of access to multimedia – data, voice, and video – on the fly, in real-time. The "shrunken" form factors for wireless devices suggest a very personal, "close to the body" form of communication with fast response times and high levels of security – imagined or real. These are attributes that 21st century business people and consumers value and are beginning to demand.

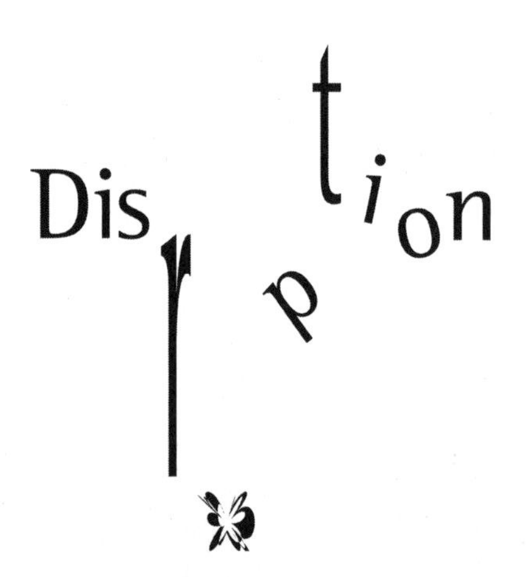

Today, for example, without even being conscious of it, wireless users are altering their life styles in the name of becoming accessible and connected, even while moving through time and space. Many are using e-mail incessantly instead of picking up a telephone. Many prefer to send one-way messages rather than tie up hours in emotional, interactive conversations. Since those who phone in cars are prone to distraction and traffic accidents, some now prefer "taking their calls" through voice recognition services and hands-free devices. Others are sending wireless messages through smart phones, Palm-connected organizers, and two-way pagers (the Research in Motion (RIM) BlackBerry is a prime example). In the white-collar world, journalists are filing stories and photographs across the Web using wirelessly connected laptops or cameras connected through their mobile phones. Mobile workers are popping palmtops and PDAs into cradles to synch their data with desktop systems – whether across the office or around the world. Sales reps that want access to corporate intranets and CRM databases are also beginning to enjoy wireless connection to their legacy systems. They can get two-way messaging, wireless push services and stock trading, easier Web access, and clearer, more intuitive wireless displays.

No matter what the desires, though, the appetite for wireless data is growing extraordinarily fast. Gartner Dataquest estimates that 36 million people will use wireless data in the U.S. by 2003, which is 12 times the number of users as last year. With the delivery of the wireless Internet, data is reaching beyond the solid vertical industries (e.g., field service, logistics tracking) that championed wireless just a few years ago to provide new paradigms for competitive advantage and productivity. Examples: more field service calls successfully completed; more inventory tracked; more customers served per hour; customer-monitored package tracking. Today, wireless data is becoming a new form of disruptive, white-collar, enterprise-oriented technology that changes business models and revenue-generating operations, permeating virtually every aspect of business life.

THE APPETITE FOR WIRELESS DATA IS GROWING EXTRAORDINARILY FAST

Consider these trends for the future:

- *Wireless data replaces guesswork:* Interactive wireless conversations happen between people and between inanimate systems. Invisible wireless devices are being implanted in rockets and under skin; in vending machines, ambulances, police cars, and taxicabs; in laptop computers, earthquake detectors and underwater reservoirs, field service trucks, utility meters, oil derricks, and nuclear power plants. Wireless data drives tractor-trailers, cash registers and automated teller machines (ATMs), warehouses and manufacturing facilities, packages and pallets, instruments, containers (which contain *Radio Frequency Identification Devices* [RFID]), marshalling yards, railroads, and executive briefcases. White-collar workers today are growing accustomed to the convenience of GPS locator systems in their cars and GPS-based asset management systems for logistics and fleet tracking. Many are using automated speech recognition devices to "drive" their wireless data calls in motion. They also browse the Internet using i-Mode and *Wireless Access Protocol* (WAP) phones for "hot sites" – from Amazon.com to CNN.com to SureTrade and the Weather Channel, even buying and selling merchandise on the wireless web. This is an age not only of automation, but also of frenetic, untethered connectedness, where information is available independent of distance.

▶▶ *Wireless devices will proliferate in all dimensions – more vendors, more form factors, more creativity.* Between 1994 and 1999 the number of mobile phones sold each year exploded from 26 million to nearly 300 million. By the end of 2003, there will be over 1 billion mobile phone users worldwide, and for the first time, mobile phones will outnumber PC and derivative devices. This explosion of devices is in response to user demand. Users want to communicate independent of their location and proximity to others. As a result, the wireless vendor landscape is very different than what we all know as the "Wintel monopoly" on the desktop, where Microsoft and Intel are the standards leaving little room for creativity or other vendors. Unconstrained by dominant forces such as these, the wireless data market is probably as remote from this world as possible. There are no standards – only creativity. Vendors are solving problems at the frontier of the user experience. Multiple chip designs, operating systems, and IT frameworks all underpin a divergent set of hardware. Although Palm, Inc. won the hearts of mobile users early with its Palm platform, phone manufacturers are shoehorning PDA-like function into the form factor of a mobile phone handset. Clothing manufacturers are inserting phones and other portable electronic gear such as MP3 players into their offerings and consumer goods manufacturers are vying for this market as well. Simultaneously, there is a design struggle between the one-size-fits-all device camp and the special purpose, form equals function camp. It seems likely that some users will want to carry a jack-of-all-trades device, sort of the Swiss Army knife for wireless communication, while others will prefer to assemble an array of function-specific devices. These discrete function devices will likely interoperate through technology known as *Bluetooth.* Bluetooth, used for wirelessly connecting devices within a short distance of each other, will enable the evolution of specific devices. These will take the form of screens (foldable and heads up which project an image a few inches in front of your eyes); mouthpieces for voice input (including voice input into a PDA where speech to text will occur); gesture gloves for mouse input; and antennae meant to be placed in a briefcase or purse.

there are no

STANDARDS

only

▸▸ *Location-based wireless services will drive mobile e-commerce.*

Mobile computing differs from desktop computing in many ways, however, the most obvious is that a mobile user's location is not fixed. This simple fact can be exploited for marketing purposes as well as for applications in both the business-to-business or business-to-employee worlds. Consider the power a pickup and delivery application can have when the home office is always aware of every truck's location or the improvement in customer satisfaction when the field service representative can accurately determine when they will arrive at your home to repair your appliance. Telematics applications for cars and trucks will use GPS data to identify the vehicle's location, speeding the arrival of a repair truck or emergency services, such as 911. A market in the several billion dollar range is being predicted for applications such as high value cargo tracking, inventory management, logistics support, and location-based billing. Suppliers of location services will utilize a variety of satellite and cellular data networks to deliver services and employ a mix of transaction charges, set-up fees, and subscriptions.

- *In the consumer space, geocentric wireless portals will drive consumption of mobile information, goods, and services.* Mobile handsets or terminals will be highly sensitive to geography as users travel to obtain goods or services instead of requiring delivery. Therefore, mobile consumers are more likely to "buy" than to "shop" willy nilly. The mobile user is likely to use a handset or dashboard wireless device to know what is physically available at nearby locations. E-coupons, therefore, could be an important driver in the mobile marketplace, especially in North America, where over 90% of all consumer buying involves travel by car.

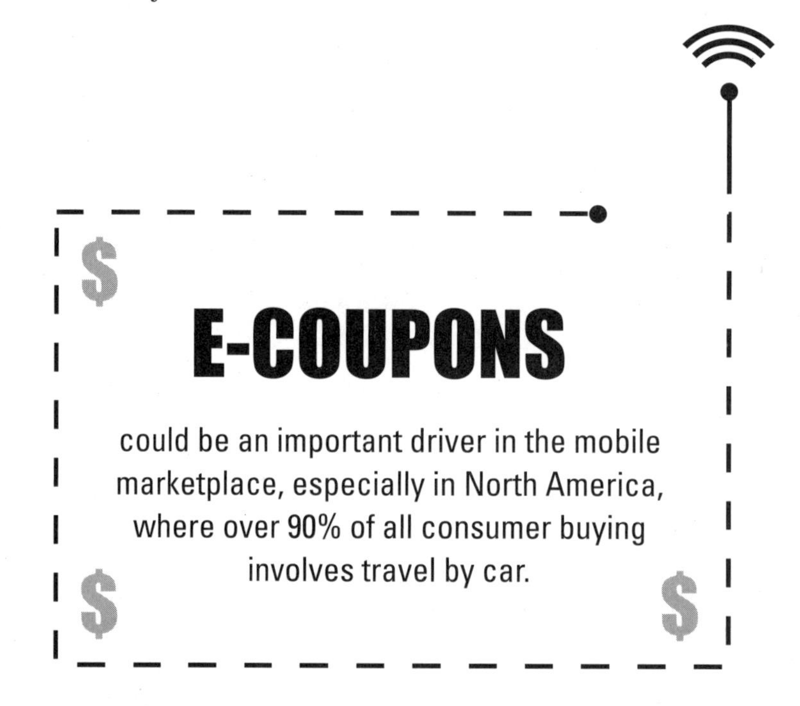

- *Wireless middleware will accelerate the integration of mobile workers and enterprise applications and Internet content/applications.* Middleware provides a "device-agnostic" translation of legacy applications into a form that wireless devices can accept. A rich and complex category of software, middleware sits between the remote user and the applications, enabling developers to build systems that mobile users require – systems that maximize available bandwidth and do not emulate the traditional desktop environment but borrow the essentials from it. In effect, middleware shields application developers from the vagaries of wireless networks and the wide variety of mobile devices, making it the key ingredient in a mobile application. Already, its use in the enterprise has transformed entire industries and business models – for example, providing superior customer service through real-time package tracking at Federal Express and its competitors. Wireless middleware provides a critical platform that simplifies the building of an application and its deployment. New mobile devices support is just one of the ways middleware provides an insurance policy that future-proofs an application.

- *Universal wireless e-mail, instant messaging, and browsing will provide a disruptive "mobile communications glue" linking the New Economy and whole segments of society together.* Internet-driven, user friendly wireless e-mail will enable all mobile users to communicate quickly and easily – especially as devices and networks become data friendly and simpler to operate. Desktop users will be able to "forward" their faxes, desktop e-mail, and other communications to a 'universal wireless Internet mailbox' on their wireless device which will be accessible wherever they are. Whole segments of the market – youth, women, special interest groups – will communicate using WAP and other Internet browsers to remain in touch. For example, Motient Corporation, formerly American Mobile Satellite, providers of the ARDIS wireless data network, now offers a plug and play Internet-enabled wireless e-mail through its eLink service, eliminating the need for proprietary data implementation. This e-Link software works with the RIM BlackBerry pagers in the Microsoft Exchange environment, allowing universal message exchange.

- T*he global mass market for mobile digital information, services, and secure transactions will exceed one billion users.* According to a recent JP Morgan Securities research report written by Paul Coster *(Out of Thin Air: Emerging Wireless Infrastructure, Software and Services*, J.P. Morgan Securities Inc., Sept 29, 2000), wireless infrastructure, software, and services will "create the largest addressable electronic market opportunity in existence and a consumer platform for conducting electronic transactions," the report states. "We believe business enterprises and wireless carriers will rapidly deploy the enabling technology necessary to wirelessly reach the global consumer (through m-commerce) and the mobile employee." Whereas "new revenue opportunities will eventually flow from the consumer, industry-wide productivity gains should immediately flow from the mobile workforce."

If we think of wireless data, then, as a journey of sorts – a journey of expectations, technology innovations, mistakes, and a vast market of emerging users and applications – we can project an end goal and describe milestones along the way. The goal is sweeping change through infrastructure improvement. In some instances, adding wireless data solutions to an existing enterprise will result in incremental improvements to efficiency, customer satisfaction, and productivity. Examples: worker hours saved, more service calls completed in a day, better, faster emergency response times; more criminals apprehended on the highway; faster package tracking, more accurate electronic inventories; customer-centric service. In other instances, wireless can result in radical new business concept innovations and strategies. Examples include over-the-air bypass of conventional credit card/ATM banking systems; cashless taxicabs and limousines; customer-focused monitoring/tracking of shipments, containers and pallets; new methods of on-the-go branding, warehousing and tagging (radio frequency identification tags), Internet-driven push services; even “wearable dating computers” that find prospects through wireless transmissions “on the spot.” Other applications include wireless stock trading; accident-scene auto claims processing; location-based retailing and directories; concierge and push services; emergency dispatch and messaging; wireless Internet buying and selling (m-Commerce); energy management and telemetry; RF-enabled security systems; GPS and direction finding; mobile transportation telemetry; wireless E-911 (smart dispatching and emergency management); Internet-based fleet management (e.g., public works); real-time mortgage and loan shopping/applications; enterprise legacy system access; streamlining of prescription delivery; extended customer resource management for sales and field forces, and many more.

It is rare for social change to be so aligned with a technological direction. Our society is moving rapidly toward total liquidity – increased mobility, shortened lead times to make decisions; increased populations of consultants, independent contractors, and freelancers; and a global economy punctuated by fits and starts of activity, including telecommuting, Internet links, air traffic delays, and transactional, real-time data. We have witnessed the growth of the small office home office (SOHO) market starting from less than 1 percent of the U.S. workforce to its current rate of 30 percent. A few years ago, information technology (IT) communities around the globe were hesitant to allow workers to work at home – fears of lax performance, disconnectedness, and lackluster management were frequently cited reasons – even though research is showing that at-home workers can be even more productive than their office-bound counterparts. Gone are the factories, where thousands of workers came together to perform work. Many world economies are morphing from manufacturing goods to performing services, and as such, the place of work is less essential and definitely less concentrated. We are, in essence, becoming highly decentralized as a working population and as such require constant communication to keep connected.

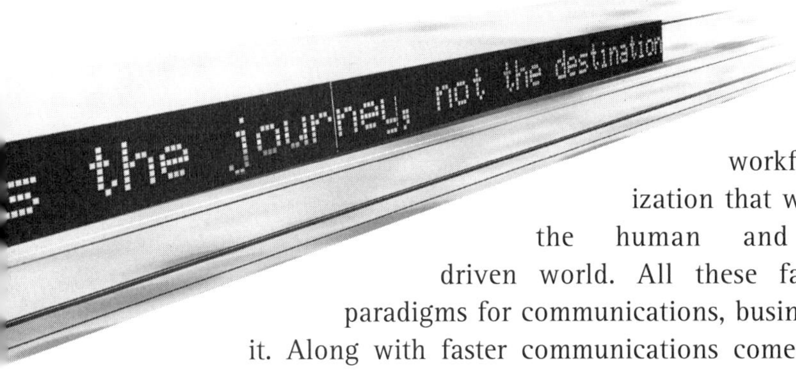

Today, the linking of enterprise resources, suppliers, and customers is becoming a natural outgrowth of the Internet and, at this time, the wireless extension of the Internet and intranet. We've seen the development of secure corporate *Virtual Private Networks* (VPN) using *Internet Protocol* (IP), the confluence of a highly paid, computer and cell-phone literate workforce, and a strong realization that wireless data can enhance the human and robotic / sensor-driven world. All these factors are driving new paradigms for communications, business, and essentially profit. Along with faster communications comes an escalating sense of urgency – a growing anxiety level that if we are disconnected for any period of time, we will "miss the action." In business, wireless data enables people to work longer hours but more flexibly, to speed up the sales or decision cycles and "get there first." In our personal lives, mobility satisfies an insatiable need to stay connected. It seems not only a natural, untethered extension of our desktop lives, but a new center of gravity. For example, real estate agents now use mobile computers before checking in at the office, not just to get housing listings but to start customer loan applications or run through credit checks. Network operators keep in touch through enterprise-level two-way messaging to receive management alerts, trouble-tickets, and network alarms in the event of after-hours failure; these device-agnostic wireless systems are selective and integrated directly with network monitoring systems. Traditional paper-driven serial processes are becoming parallelized, compressed in time and space through the use of wireless data – creating 'nonlinear' applications. In writer and business analyst Gary Hamel's words:

> "In a nonlinear world," Hamel writes in his book Leading the Revolution (Harvard Business School Press, 1997), "only nonlinear ideas will create new wealth. Most companies long ago reached the point of diminishing returns in their incremental improvement programs. Continuous improvement is an industrial age concept, and while it is better than no improvement at all, it is of marginal value in the age of revolution. Radical, nonlinear innovation is the only way to escape the ruthless hypercompetition that has been hammering down margins in industry after industry. Nonlinear innovation requires a company to escape the shackles of precedent and imagine entirely novel solutions to customer needs."

Call this brand of reinvention a modern-age "flight of ideas," as 18th century novelist Laurence Sterne would put it. But nonlinear innovation is a hallmark of the New Economy and wireless data solutions in particular. Even the classic analog RF signature, the wavy line – is now broken up into discontiguous digital codes (i.e., *Code Division Multiple Access* [CDMA]) sent over the air, suggesting a new reorganization of corporate assets, knowledge, and processes. Like e-mail, wireless data is disruptive because it promises to reconfigure business processes by providing something simpler, cheaper, smaller, and easier to use than existing technology. And consistent with other disruptive technologies, wireless data can be more cumbersome and a "worse" alternative than mainstream buyers want to accept – for now. This means that wireless data has merit but not universal applicability. It doesn't mean that businesses can ignore wireless data waiting for all technological improvements to be made. Wireless data technology will vastly improve in many areas over the next few short years, such as issues of network coverage, security, bandwidth optimization, networking standards, diverse number of players in the value chain, availability of enterprise and compelling consumer applications, device interface, clarity and convenience, and much more. But because wireless in all its forms (e.g., data, video, text, graphics, et cetera) is the journey, not the destination, its implication for weaving a new fabric across existing ways of communicating and transacting business and pleasure – is profound.

This book will chronicle the journey, describing how untethering into wireless can become an occasion for innovative, nonlinear, disruptive business thinking. It will describe through real-world case studies exactly how wireless data has changed models for business and profit generation in key industries. It will show how pioneer and mainstream adopters have used wireless networks and middleware to reengineer an enterprise solution, radically alter a business, or simply improve one.

This book also portrays the actual "experience" of using wireless data on the road and in the enterprise. It will show managers and users what it's like to imagine a new market or usage for wireless, implement a solution, rethink a process, make a connection, download an application, and troubleshoot it – both in consumer and enterprise spaces. It will explore the concept of design tradeoffs vs. user tradeoffs – and discuss the issues of security, network type, speed, price, proprietary vs. "off the shelf" solutions, build vs. buy middleware. It will help you develop an integration plan for using wireless across the enterprise, linking personal, local, and wide area employees in the wireless data space.

Fantastic new wireless devices and networks are around the corner, and it's clear that a whole new generation of technologies will enable the next step in globalizing the mobile Internet. Wireless data isn't easy. It has proven to be a slower, less amenable, less market friendly phenomenon than innovators originally anticipated. But because of its disruptive, revolutionary potential – through infrastructure improvement, both communications and business – the long-term impact of wireless will be enormous. The early rewards of this are already being seen today.

Read on to find out how mobile she devils and wireless data have become one with history.

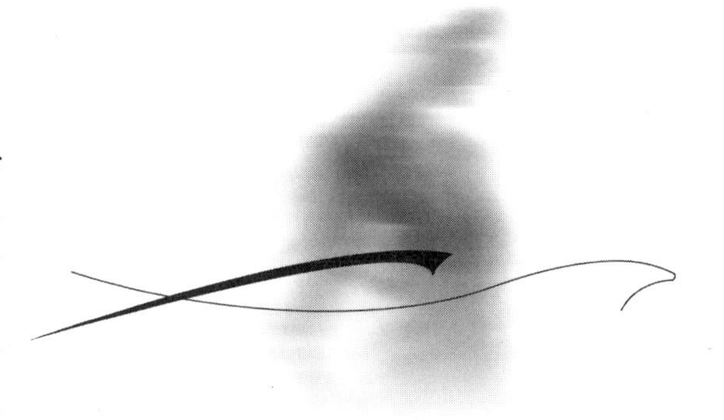

secret communications system

2 NEEDED AND DO-ABLE: A WIRELESS DATA HISTORY

HEDY LAMARR, A GERMAN ACTRESS WHO PLAYED the seductress and she devil 'Delilah' in *Samson & Delilah* with Victor Mature, was one of the important figures in mobile data development. She co-invented the digital signaling technology, *Code Division Multiple Access* (CDMA) during World War II, which was ultimately to become the most popular North American standard for wireless communication.

Lamarr, who was married to a German munitions manufacturer, Fritz Mandell, fled the country with knowledge of Nazi military secrets and became an actress in London with filmmaker Louis B. Mayer. Later, in Hollywood she collaborated with George Antheil on an ingenious, secure military signaling code she dubbed "Secret Communications System." Lamarr and Antheil won a patent for it, conceiving it as a guidance system for Allied torpedoes.

The original concept for CDMA encoding consisted of using two paper rolls (similar to those used in player pianos), punched with an identical pattern of random holes, according to Bob Stoffels, in his excellent article "The CDMA Film Noir," *America's Network*, June 1 1998, page 22.

One of the rolls would control the transmitter on the submarine; the other would be launched with the receiver on the torpedo. The gorgeous Lamarr never saw her invention made practical – the Navy deemed it too cumbersome. Nonetheless, by the late 1950s, Navy contractors were able to take advantage of advanced electronics to launch and synchronize the first military CDMA signaling systems. In the late 1960s, a brilliant electrical engineer named Irwin Jacobs began perfecting CDMA satellite technology in the form of VideoCypher, a scrambling system designed for satellite *C-band communications* from sky to phones on the ground. Jacobs, who became the chairman of Linkabit and later, Qualcomm, commercialized CDMA as an international voice and data standard for digital mobile networks. Today, Americans know Qualcomm and CDMA as a great success story – one example of how a real communications need, coupled with serendipity, genius, and a steady stream of technology improvements – converged over time to produce a commercial success.

The Lamarr-Jacobs CDMA story isn't unique. The whole concept of wireless data – at the very least, communication at a distance – goes back as early as the Bible if you count the fanning of fire and brimstone as the first step in the journey. The Bible tells how Moses led the children of Israel out of Egypt following a column of fire and smoke, and the Greek poet Aeschylus (525 – 426 BC) described the use of the first optical telegraphy – fire by night, smoke and mirrors by day – in his epic poem, *The Agamemnon.*

Historically, wireless innovation has always satisfied two basic conditions. The first is a real world "need" – e.g., the need to send secure signals from one tribe to the other during wartime, or to communicate news or keep track of people and freight in peacetime. The second requirement is technology capability – enough horsepower and signaling to meet the designated need and be "do-able" – as is practical and cost-effective. Those two conditions, "needed and do-able," drive wireless data networks even today. From primitive communication to personal messaging, from Allied torpedoes to the invention of wideband digital CDMA, wireless has always served a purpose. Each technology was created in the context of a society eager for a better way to impart information, maintain privacy, and communicate anywhere on the fly.

How Needs Drive Technologies — and Vice Versa

Think of a system of needs and technologies as spiraling, gyre-like, and interactive. By 1971, the Bell System in the U.S. had submitted a cellular radio scheme based on frequency re-use to FCC, but six more years would pass before FCC allowed AT&T to start a trial. Without the technological capability, for example, the need to communicate at a distance is defeated. You cannot communicate (save for telepathy) with a mobile worker or a customer hopping planes and trains all day unless you find a way of securing a channel, using a network, and reaching that person who carries a mobile communications device. Conversely, without sufficient need, technology is senseless and superfluous (even if it's ingenious). The patent museums are littered with great gadgets that never caught on because a real need wasn't satisfied; the devices were shelved, abandoned, or eventually, transformed and tweaked to meet a different cultural or business need.

In this balancing act, applications (e.g., the outgrowth of needs) and technologies act on each other in a bi-directional flow. In other words, inventors create devices and applications that respond to the needs of society; and then technologies refocus and transform societies, enabling users to "discover" new needs.

"Needs" and "Technology": Pigeons to Ponies to PDAs

As an example of how "needs" push "technology," the Pony Express solved the problem of communicating faster by organizing the existing technology into a cohesive delivery system leveraged on the true grit of riders and their horses. Obviously a marked improvement over other techniques, the Pony Express sparked the application of "what else could I do if I could communicate faster?" This completed one cycle by having "technology" push "applications." Further spins of "need" and "technology" created a national train network that created a mechanized improvement over the John Henry-esque Pony Express and when people demanded even faster delivery, Samuel F. B. Morse delivered the telegraph.

It is important to note that technologies solve a certain class of problem, typically presented as the "needs frontier" of the day. They have inherent value in their time, meaning that waiting for the ultimate technology is a fool's gambit. Today we would not try to solve the problem of how to deliver a message in less than several months, yet in the day of the Pony Express, this was a fundamental achievement. In the light of today's technology, the telegraph is tucked into the back of a communications museum as will be today's technology probably only 10 or 20 years from now. This doesn't mean that today's technology is unable to solve today's problems and that as a society, that we should wait for a future technology to solve today's problems. Quite the converse, today's technology is well suited to solve a wide range of commercial applications and should be applied as such.

Virtually every successful communications method has responded to the unique needs of the era for which it was created. Think about the first telegraphs, the Marconi radio, the landline telephone, the first pagers, the first wireless cell phones, and PDAs. In each case, needs and available technologies and tools dictated a solution.

Encoding was an early example. The concept of using signals to represent words and messages goes back thousands of years. A *network* to carry the signals started with smoke, light, and chains of human voices. It progressed to electricity, wires, and then air (the medium of propagation for radio waves). *Communications devices* moved from cumbersome wooden poles (see the water telegraph on this page and the next) and mirrors ultimately to metal keys punching out Morse code, large cumbersome handsets, and finally bulky radios with vacuum tubes and then transistorized radios that fit into the palm of one's hand. In the last decade, microprocessors and digital switching systems enabled new forms of wireless encoding, compressing information into individual, secure channels.

New Communications Methods Change the Way We Work and Live

Moreover, technologies changed social expectations. Today, for example, people expect voices, messages, files, and information to be delivered in real-time, anywhere, 'on the fly,' 24 X 7. Wireless devices are mobile, "close to the body," available, waking and sleeping. In short: Wireless data has changed personal habits, social ways of communicating, means of corporate organization, and decision making; ways of reaching the customer; and, bottom line, ways of making a buck.

Consider:

- Wireless devices are now small enough in size, weight, and portability to be attractive to mobile users.

- The machine interface for wireless – the size of the screen, for example, along with the development of durable batteries with longer life, the availability of *qwerty* keyboards, thumbwheels, touch screens and other devices – have increased the practicality of devices and made them easier to use.

- Wireless networks now have sufficient coverage to enable personalized conversations, data transmission with comparatively low latencies (delays), and reusable frequencies, making networks more efficient.

- Wireless applications and content are becoming richer each day. Witness the flowering of wireless connections to the Internet; wireless links to corporate intranets across firewalls, the expansion of applications ensuring not only mobile employee communication but real-time customer-centered communication. Content is becoming more robust and specialized for mobile users.

- Gateway software (e.g. middleware) configures wireless content and reformats data for mobile devices. In enterprise networks, gateways translate content from 'legacy' information systems (e.g., "wired content") into forms wireless devices can accept.

Who would have thought that the invention of paging networks and the car phone would ultimately result in teenagers being 'glued' to their pagers and phones in virtually 'always on' fashion? Who would have imagined that wireless gateways could one day enable mobile field and sales forces to transmit their whereabouts and get access to data, turning around quotes and contracts directly to customers via mobile devices?

Imagine the early desktop experiments in e-mail and messaging in primitive ASCII and DOS formats. Eventually we had powerful, two-way text messaging on phones and PDAs, plus the Windows® interface, the explosion of Internet traffic, streaming videos, location-sensitive directories, downloaded NAPSTER music, cartoons, punk communication, GPS mapping, and direct links to wireless terminals. On Wall Street, "24 X 7" mobile stock trading is the reality. The capability to get quotes wirelessly on a pager or palmtop, to make transactions in real-time, to set alerts and alarms, to avert disasters – and to trade while in taxi cabs, trains, or after hours, accessing worldwide markets in Tokyo, Paris and New York – has become a searing investor requirement.

No one could have anticipated these events several thousands of years ago or for that matter, even a decade ago. But streams flow into rivers, and a basic structure for communication began to unfold from which today's devices, networks, and expectations were ultimately derived.

In the case of wireless, technology capabilities had to be satisfied at four major functional levels: the *user device* (e.g., a portable handheld terminal); *the network*; the *application/content*; and the *gateway* (software or middleware enabling the translation of signals from ground-based systems to those over the air). Even in primitive forms, all these components had to be in place in a "do-able" fashion in order for the technology to thrive. When needs and applications were compelling, early adopters took the necessary technology risks.

Now let's go back to wireless history, to see how these separate streams of communication – telegraphy, radio, and mobile telephony – evolved and converged to produce a structure for wireless data today.

The Power of Convergence: Telegraphy, Telephony & Radio

At first, the concept of wireless was primarily 'visual' – signals, fire and brimstone, smoke and mirrors, or a chain of human voices – to carry messages across distances. Later, the concept of 'network' evolved to include optics and energy – electricity, wiring, and then "ether" – the medium of propagation (air) for radio waves.

Telegraphy was the first great tributary of communications. In the 4th Century BC, Aeneas the Tactic described a visual concept for communicating at a distance: the water-driven telegraph. Invented by the Cartaginese, this ingenious system used two cylindrical vases (a transmitter and receiver), which were perfectly identical and placed on two distant hills. Filled with water, the vases had a floating vertical pole at the center with conventional signs attached to it. To communicate, soldiers raised or lowered the pole by emptying or pumping water in the vases to the desired point. Flags or torches indicated the start and completion of transmission.

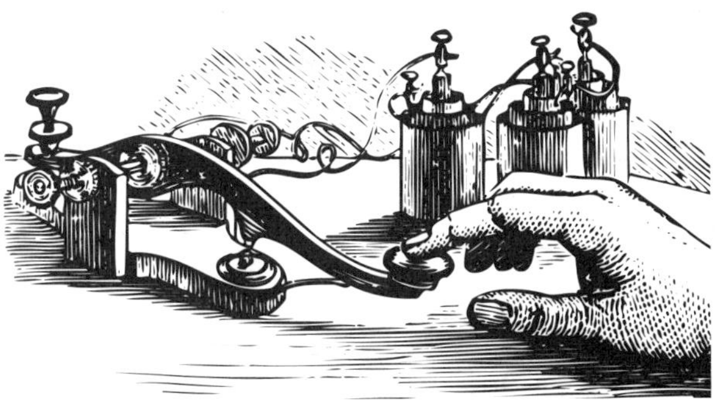

In France the first true optical telegraph emerged in 1792 AD, when Claude Chappe (1763-1805) and his brother Ignace demonstrated their invention to the French legislative assembly, which therein marked it for official use in warfare. Chappe's telegraph generated signals based on different positions created by three wooden interlinked arms. The central regulator was longer than the other two indicators (known as "wings") and rotated at the top of a vertical fixed pole. The two lateral arms, rotating freely around the center, had displacements of 45 degrees; and different positions of the arms could transmit a rich vocabulary of up to 8500 words. (Two signals were required per word.) The first telegram sent by the Chappe system announced a French victory over the Austrians on November 30, 1794; subsequently the Chappe telegraph spread throughout the territories of Europe, and became a common means for nations and warring states to communicate encoded messages.

Shortly after Chappe's inventions, American inventor and painter Samuel Morse (1791-1872) was the first to invent a practical electric telegraph in his painting atelier. His first receiving instrument was constructed on a 'canvas stretcher' frame. A wooden clock motor provided the power to move a paper tape under a pen; in actuality, an electromagnet moved the pen, and was driven from the telegraph line. The canvas stretcher provided a frame to support these devices, and on May 24th, 1844, Morse sent the famous words from the Bible, *What hath God Wrought!* on his telegraph from the U.S. Capitol Building in Washington, DC, to the B70 Railroad Depot in Baltimore, Maryland. Therein began one of the most important chapters in communications history; and a means of linking a vast continent about to experience rapid industrial expansion and Civil War. (However, telegraphy is not actually 'wireless,' even though it's about signaling and data. Actually, electrical telegraphy relies on wireline transmission – if someone cuts the wires, the transmissions end. Still, the innovations in signaling and networks were vital for later wireless development; therefore telegraphy is still considered a mainstream source of components and ideas for the wireless signaling networks of today.)

Paralleling telegraphy developments were discoveries in electromagnetic science and radio. In 1887 German Heinrich Hertz used periodic currents at very high frequency to demonstrate the existence of electromagnetic waves. Hertz constructed the first spark-gap transmitter, a device that generated radio waves from an electrical spark. Today, few people know of this particular invention but his name is mentioned millions of times everyday around the world, since it is used to measure frequency, as in microprocessor speed, e.g., 1.4 GHz (giga Hertz). Hertz could not have foreseen the microprocessor, however, his work on frequency proved to be fundamental to future technology breakthroughs.

The Beginnings of Wireless

In September 1895, self-taught Bolognese wunderkind Guglielmo Marconi (*see Figure 2.1: "A Brief History of Wireless"*) began performing simple radio experiments. Marconi showed he could send signals by using electromagnetic waves to connect a transmitting and receiving antenna. Though popular wisdom at the time held that electromagnetic waves could only be transmitted in a straight line, and would be disrupted by the curvature of the earth, Marconi disproved the theory. He placed his transmitter near his house and the receiver three kilometers away behind a hill. Marconi's servant, Mignani, was instructed to fire a rifle shot

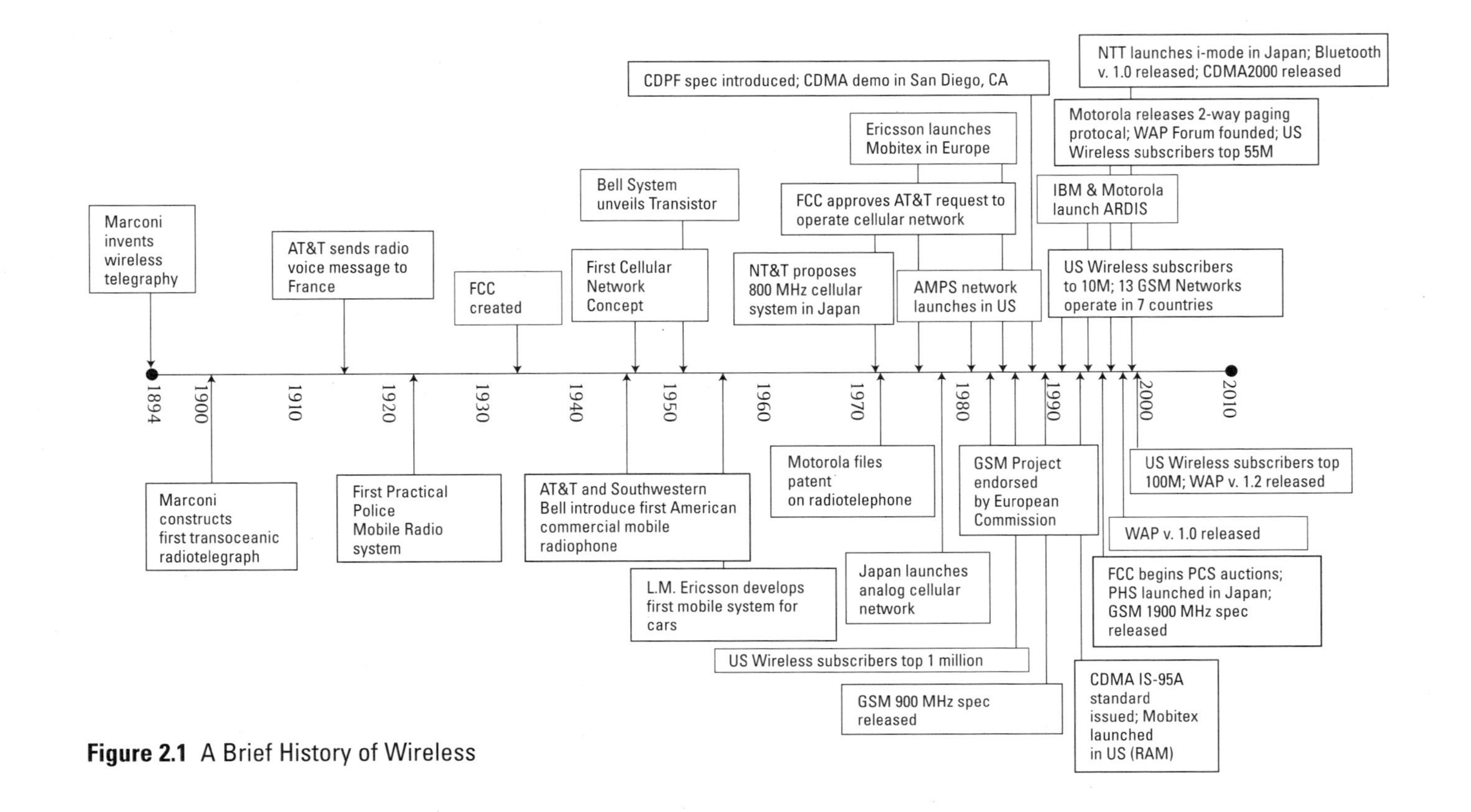

Figure 2.1 A Brief History of Wireless

when the signal was received. When Mignani fired his gun, he signaled that three dots of the letter "S" of Morse code had traveled through space unimpeded by geography or earth's curvature.

Though Marconi found little enthusiasm for his invention in Italy, the British Ministry of Posts took up his cause in 1897, giving him money and technicians, and soon radiotelegraphy was made practical. Radiotelegraphs used radio waves rather than wire telegraph lines to send and receive messages; thus they were the true precursors to wireless data of today. By 1901, Marconi's team had constructed the first transoceanic telegraph between Newfoundland, Canada and Poldhu, England. In 1905, the first wireless distress signal was sent via Marconi radio using Morse code – a series of long and short signals developed to represent the letters of the alphabet. This also shows that two generations of technology can be blended seamlessly: Morse's code, which was originally designed for telegraph transmission, and Marconi's wireless network. This pattern of re-using technology is prevalent in today's wireless data networks.

Wireless received another boost when American inventor Lee De Forest helped refine vacuum tube technology in the early 1900s, enabling amplified wireless signals to send voice transmissions. In 1915 the American Telephone & Telegraph Company (AT&T) sent a voice message by radio between the United States and France, followed two years later by Bell System engineers demonstrating two-way radiotelephone transmission from airplane to ground. Wireless voice was now in full swing. In the early 1920s, the first practical police mobile radio systems began operating at 2 MHz, just above the present AM broadcast radio band. During the Depression, civil emergency and law enforcement authorities commonly used small two-way radio transmitters as a simple application of the new technology.

Advances in Mobile Telephony and Military Radio

The concept for digital wireless and cellular voice and data is rooted in the late 1930s and 40s, when commercial mobile telephony and military radio systems reached production quality. At that time, the U.S. military began experimenting with radio frequency signal encryption for data, using it to send battle plans across enemy lines and naval fleets ship to shore. Lamarr and Antheil began work on the "secret" CDMA signal guidance systems that would eventually find a true home in wireless phones. Meanwhile, in 1934, the U.S. Congress created the Federal Communications Commission – a regulatory body designed to manage landline interstate telephone business as well as radio spectrum – the "ether" required to make radio networks work. The Commission began parceling out spectrum parsimoniously at first, giving priority to emergency services, government agencies, utilities, and other groups deemed helpful to people. After World War II the FCC began to allocate radiotelephone channels for commercial use. In June 1946 in St. Louis, American Telephone and Telegraph and Southwestern Bell introduced the first American commercial mobile radiotelephone, the precursor to mobile wireless today.

Early Mobile Radio Telephone Systems (MTS) were plagued with cross-channel interference, forcing the carrier to restrict usage and the number of channels available for voice calls. (Channels are important because they relate directly to how many communications sessions can be maintained by a radio tower at any one point in time – and therefore directly affect the costs of transmission.) Therein began an ongoing struggle for spectrum that has lasted right up until the present day, wherein carriers depend on technology breakthroughs – boosting radio signal efficiency, for example – to make better use of the crowded bandwidth available.

Roots of Cellular: MTS Voice Systems

The earliest MTS system for voice utilized a honeycomb design prefiguring cellular systems. The honeycomb consisted of one central tower and five receivers. Combining signals from one or more receivers into a unified signal, the tower amplified them and sent them onto a toll switchboard. This allowed roaming from one city neighborhood to the next. Voice communications were half-duplex; one party talked at a time by pushing a handset button to talk, then releasing it to listen. In MTS the base station frequency and mobile handset frequency were offset by 5 KHz, so that eavesdroppers could only hear one side of a conversation (an early attempt at privacy); a caller searched manually for an unused frequency before placing a call. The systems were fairly reliable, but users complained of too few available channels.

The early MTS systems built the foundation for many cellular developments in years to come. Bell Laboratories scientist D.H. Ring first articulated the cellular network concept with coauthor W.R. Young in 1947. Their technical memorandum outlined the essential elements of a cellular network – including a network of small geographic parcels or "cells," a low powered transmitter in each, a central switch, and frequency reuse. Ring and Young predicted that a wireless system could optimize frequency reuse in small cells, thereby creating a blanket of wireless coverage. This blueprint for cellular departed from existing mobile telephone systems that required the call to be transferred from zone to zone as the mobile unit traveled. Instead, cellular switches the frequency it is placed on as the mobile unit moves (since frequencies change from cell to cell); hence frequency reuse became a trademark of the more efficient cellular systems.

In 1948 the Bell system unveiled the first transistor; this was a breakthrough that enabled radios to "shrink" down in size to small, handheld radio transceivers, enabling civil authorities to communicate easily with each other. Public two-way radios with several frequency options became available. In Stockholm, the first phone-equipped cars took to the road by the mid 1950s when pioneer Swedish telecom company L.M. Ericsson developed a mobile phone apparatus consisting of a receiver, transmitter, and logic unit mounted in the boot (trunk) of the car. The system voraciously consumed battery power and was costly and impractical at first, but it set the stage for cellular car phones. By the mid 1960s, new wireless equipment using transistors were brought onto the market, making analog mobile telephony, and ultimately, digital voice and mobile data feasible.

Spread of Analog Cellular

Analog cellular systems became widespread in Europe in years after the assassination of John F. Kennedy and the Vietnam War. In 1967 Nippon Telegraph & Telephone company, the Japanese equivalent of AT&T, proposed a nationwide cellular system at 800 MHz in Japan. In January 1969 – the year Neil Armstrong landed on the moon – the Bell System made commercial cellular radio operational by employing frequency reuse for the first time on trains. Delighted Metroliner passengers could make cellular calls from payphones while traveling between New York City and Washington, DC.

By 1971, the Bell System in the U.S. had submitted to the FCC a cellular radio plan for full-scale commercial service to consumers, but six more years would pass before the FCC allowed AT&T to start a trial. A few years later, on October 17, 1973, Dr. Martin Cooper of Motorola filed a patent entitled "radiotelephone system," which outlined Motorola's first ideas for cellular radio. Cooper set up a base station in New York with the first working prototype of a cellular telephone and called over to his rivals at Bell Labs in New Jersey. Both Bell Labs and Motorola were fiercely competitive in the 1960s and early 70s, trying to make cellular communications technology practical (it was not until 1977 that the FCC approved AT&T's request to operate a cellular system). And despite incredible demand, it took 30 years for cellular to go commercial in the United States from the mobile telephone system's first introduction. The delay came, in part, through bureaucratic bungling, but radio common carriers, which provided conventional wireless phone service in competition with AT&T, also played a part. Carriers like American Radio Telephone Service, and suppliers to them like Motorola, which catered to the radio communications needs of companies ranging from taxicabs to marine and naval ship-to-shore, feared the Bell system would dominate cellular radio if private companies weren't allowed to compete equally. The radio carriers fought for equal access in court and wanted the FCC to ensure an open market.

At America's Bicentennial, 1976, only 44,000 Bell subscribers had AT&T mobile telephone systems, and an additional 20,000 people sat on five to ten year waiting lists. Radio common carriers had roughly 80,000 units deployed in the American market. Spectrum was a primary deterrent. As late as 1978 the Bell System, independent carriers, and non-wireline carriers divided just 54 channels nationwide for wireless; and this compares to 666 channels that the first analog cellular systems would require for operation. Most telcos gave scant attention to mobile services, focusing instead on delivering basic telephone service to the masses. The 1980s at last saw the deployment of primary analog cellular technology for voice, *Advanced Mobile Phone Service* (AMPS) operating in North America at 800 MHz. Worldwide commercial AMPS deployment followed quickly, and several competing standards for digital voice and data ultimately emerged (see Figure 2.1).

It's interesting to note that virtually all new and disruptive communications technologies have chequered histories marked by regulatory battles and long delays in adoption. Facsimile, for example, was developed in Bell Labs in the 1920's, but commercial fax was not launched until the 1960s, and mainstream use came to be in the 1980s. Wireless has followed a similar path: at first inventions took decades to become commercial realities; but now the turn-around time for commercialization can be a matter of months.

The Push for Digital Voice and Data

The true wireless revolution for voice and data did not begin until the advent of low-cost microprocessors and digital switching (1980s - present day). Ironically, the North American Bell System, while building the finest landline telephony system in the world, never seemed truly committed to mobile voice, much less mobile data, until the mid 1990s. In addition, federal regulations hindered development of wireless services by tightly regulating spectrum. By contrast, in Europe and Japan, where governments regulated their state-run telephone companies but permitted rapid building of wireless networks, both analog and digital mobile telephony came in sooner and less expensively, making the networks attractive to millions of subscribers.

Wireless messaging grew fast in the paging world, where the British Post Office had invented POCSAG, a paging standard to support very short messages on the original numeric pagers at slow speeds (512 bps to 2400 bps). Motorola came out with a competitive, faster network infrastructure for paging called FLEX (one-way paging, a synchronous protocol designed in 1993), which became virtually ubiquitous. FLEX paging networks offer five times the network capacity as POCSAG, support longer messages, and are supported in 30 countries and by most paging carriers in the U.S. By 1997 Motorola had released its more powerful REFLEX infrastructure, a two-way paging protocol offering varying inbound and outbound channel speeds (up to 19.2 Kbps), support for alphanumeric paging, frequency reuse, guaranteed message delivery and response, among many enhanced services. REFLEX and inFLEXion (voice paging protocols) are rapidly replacing the old POCSAG standard in North America, and paging devices are now used actively in over-the-air wireless telemetry applications, among them utility meter reading, automobile paging systems, industrial controls, automated vending machine inventory control, crime fighting and monitoring devices, and digital image transfer.

The First Wireless Data Networks

Concurrent with paging developments, telecom giants began developing 'data only' wide area specialized networks for mobile users. In 1985 the Swedish giant Ericsson launched its mobile packet network, Mobitex, in Europe; and Motorola and IBM invented a comparable standard, DataTAC, for wireless data services in the U.S. and abroad. Further, standards were promulgated for *Short Messaging System* (SMS), a protocol used primarily for two-way dispatch networks aimed at fleets of vehicles and now used by practically anyone who has a mobile phone in Europe to send and receive short messages. These specialized networks appealed principally to business and vertical markets, employing handheld or vehicle-mounted wireless terminals, custom middleware, and specialized applications. The networking technology took advantage of wireless "packet" or 'always on' transmissions similar to X.25 (a common wireline packet protocol). *Packet switching* means the networks do not depend, as circuit-switched networks do, on securing an open channel or reserving it for the exclusive use of the sender. Rather, each packet of data is individually addressed and routed to its final destination, increasing network efficiency and lowering cost (see Chapter 7 on "Wireless Networks" to learn more about packet vs. *circuit-switched* technology). By contrast, circuit-switched data, the most common form of data in current voice cellular systems (e.g., CDMA or *Time Division Multiple Access* [TDMA] networks), assigns each data transmission to individual radio frequencies. These systems reserve frequency or time slots for an individual transmission and don't allow multiple users to contend for spectrum once it is allocated.

Why are these distinctions important? Wireless networks have to deliver data efficiently, with high performance and comparatively low cost. Early investments in wireless data services had to be justified – mostly by return on investment (ROI) based on tangible and some intangible benefits. One of those benefits was transmitting data over the air on an "anytime, anywhere" basis. Packet networks, in particular, offered mobile users a new level of wireless "over the air" performance - especially for limited (narrowband) mobility applications, such as dispatch and field service. Ericsson's Mobitex network, for example, could operate at 8 Kbps, with latencies of no more than 10 seconds, which was deemed acceptable for many commercial applications. Motorola and IBM, meanwhile, began collaborating on a separate optimized data technology, DataTAC, for IBM's mobile field force. The original technology (called MDC) operated at 4.8 Kbps; IBM installed a network linking 20,000 field service employees using MDC and a communications device that was large and bulky, which earned it the nickname of the "brick."

Eventually, Mobitex in the U.S. became RAM Mobile Data, which spread nationally in 1993, and was eventually absorbed by BellSouth as BellSouth Wireless Data (now Cingular Interactive). At roughly the same interval, IBM and Motorola's DataTAC was transformed into a competing, nationwide wireless data network, called *Advanced Radio Data Information System* (ARDIS). IBM's field force first used ARDIS for dynamic dispatch, parts ordering, entitlement checks, and technical assistance from repair centers. Ultimately, ARDIS became the chief network for American Mobile Satellite Corp (renamed Motient in 2000), deployed throughout North America to major corporate accounts, among them United Parcel Service. Both RAM and ARDIS set basic standards for nationwide mobile data coverage that have yet to be surpassed.

Digital Standards for Cellular Proliferate

Meanwhile, in the cellular world, new standards for digital voice and data networks also emerged. By the early-1990s, providers were beginning to implement digital networks (many of them considered "overlays" to existing analog networks). The most popular digital standards included D-AMPs (digital AMPS), *Pacific Digital Cellular* (PDC, a Japanese Standard), and *Global System for Mobile* (GSM), the latter standard spreading widely throughout Europe and the rest of the world. In the U.S., Asia, and parts of South America, CDMA, championed by Qualcomm, Sprint, Bell Atlantic (now Verizon) among many others, became competitive against TDMA, a standard used by AT&T Wireless. Further, a new type of packet data network, known as *Cellular Digital Packet Data* (CDPD), was implemented on top of the existing analog networks to enable national wireless carriers to offer reliable data services. Though comparatively late out of the starting gate (1989), CDPD rapidly became popular for wireless data networking in vertical markets, including dispatch, public safety and emergency services, and field service. Today CDPD is the primary data mobility network for police and state government agencies; it competes against the other specialized networks, Cingular Interactive and Motient, in the commercial realm.

Wireless Data in Our Time

As digital cellular networks in the U.S. continued to evolve, they became quickly overburdened – virtually out of capacity in the more densely populated cities. By the mid-1990s, the Federal Communications Commission (FCC) began auctioning spectrum for "next generation" digital in the newly designated "Personal Communications Services" PCS band (1900 MHz). This opened a new chapter in wireless competition. Fearing that they could be boxed-out of offering services if they didn't own sufficient radio spectrum, companies fiercely competed to acquire spectrum. This pushed the prices into the speculative range and put enormous pressure on carriers to build subscribership – causing bankruptcy for many smaller PCS start-up companies, as well as halting infrastructure build-outs. Many competing wireless carriers supporting different digital standards – among them, GSM, CDMA, and TDMA – fought for subscribers and turf. Some large wholesale and retail providers were gobbled up (e.g., PrimeCo) or dropped out of the race (in a stunning turn of events, NextWave Telecom, which had its licenses taken away by the FCC after defaulting on its license payments, was given those licenses back on June 22, 2001 when the U.S. Court of Appeals for the District of Columbia ruled that the U.S. government violated bankruptcy law when it revoked wireless PCS licenses obtained by the company. NextWave has since raised $200 Million to bring itself out of bankruptcy and contract with Lucent to build a 3G digital wireless network, based on CDMA). Many were forced to cut prices to nearly bare bones levels because of steep metropolitan competition. (Meanwhile, many rural areas were neglected; and digital coverage is still spotty in vast rural and suburban stretches of the country leaving room for satellite-based wireless service providers.)

Today, the digital wireless turf wars continue. But in the last few years a wave of industry consolidations and mergers has left fewer, albeit stronger competitors. Some of them are planning to roll out even more robust 3G wireless voice and data services with broadband capabilities supporting large file transfer, color graphics, and multimedia. Though time frames are uncertain, companies such as Verizon, AT&T Wireless, and Sprint are planning trials and implementations of broadband wireless services for enterprise and consumers (supporting up to 2 Mbps) within the next few years. Many of these wireless carriers are pushing mobile Internet services, sometimes known as "wireless web" – news, chat, location-based services, directories, and access to selected Web and mobile e-commerce sites. Meanwhile, Cingular Interactive continues to operate the Mobitex nationwide network; Sears Product Repair Services is one of its biggest customers. Motient Corporation operates ARDIS for many enterprises. CDPD and *Specialized Mobile Radio networks* (SMRs – Nextel is an example) are used widely in service businesses, public safety and government, emergency dispatch, and some field force applications. (See the next chapter for details on real world case studies, such as Sears and others.)

The Drivers for Wireless Today

Strong "vertical markets" for wireless mobile data initially drove early deployments in the 1990s and remain the most robust center for data today. Blue collar verticals – manufacturing concerns, utilities, warehousing, field service, dispatch, package tracking – were among the earliest adopters of digital wireless data because they could show a high return on investment (ROI). The precedent was set in the 1980s, when Federal Express (FedEx) launched a groundbreaking, proprietary radio data system, creating the first wireless real-time tracking system. This gave FedEx an enormous lead over other shipping companies endearing them to customers, because FedEx could provide a much faster, better-documented service. Competitors were forced to revamp the way they did business. UPS quickly jumped into wireless data for package tracking in the early 1990s, and the company now is deploying its second generation system. Similarly, FedEx Ground, which used to be called RPS (Roadway Package System), is now rolling out an application running on the Mobitex-based packet data network from Cingular Interactive, with enterprise middleware solutions from Broadbeam Corp. that reach 5000 FedEx Ground employees.

Sears was another wireless data pioneer. Its Sears Product Repair Services division began reengineering customer service and field force repair in the late 1980s. The company originally found that service technicians made prodigious numbers of phone calls from pay phones back to the home office to order products and parts, check stock, and revise repair schedules. This took technicians approximately 30 minutes each day – time they could otherwise spend making customer service calls. Other inefficiencies were also identified and quantified. Customers would sometimes cancel calls while technicians were already enroute; or they would be unavailable to answer the door. All order forms and service bills had to be filled out manually along with other administrative tasks. It was concluded these inefficiencies could cost justify a dynamic dispatch where routes could change according to calls and cancellations during the day.

Sears initially selected a wireless data system consisting of a ruggedized handheld device from IBM, communications network by ARDIS, and a DOS-based field service application designed by the Sears information systems team. But after initial tests and IBM's exit from the ruggedized marketplace, Sears Product Repair Services adopted a ruggedized handheld PC (Itronix XC-6000) with a portable printer and wireless capability. Productivity benefits were evident even in the initial field trials. Many technicians were able to boost service call rate from 6 to 8 per day – a critical increase because on average 6 calls per day are covered under warranty provisions. Call completions on the first call also increased dramatically – an important metric because callbacks to finish service calls are expensive. Ultimately, phone calls from technicians to order parts or confirm information went down by 50%. Wireless data communications between the dispatch center and technicians became routine, automated, and effortless. Sears' management began increasing its profitability, setting a precedent for moneymaking service operations throughout the retail industry.

Sears and FedEx are good examples of enterprise wireless data pioneers. Other vertical applications for wireless flourished early – some of them driven by local area networking wireless requirements, some wide area. Inventory control, load tracking, and warehousing became important applications. Manufacturers began experimenting with short-range bar code scanning/inventory tracking networks – wireless *Personal Area Networks* (PAN) linking to wireless LANs and even *Wide Area Networks* (WAN). Field service and sales applications proliferated; many trucking fleets, freight companies, and service organizations used the early networks to track freight, "tag" loads, and give truckers instantaneous messages about routing. Meanwhile, wireless LANs proliferated in hospitals and universities, linking diagnostic equipment, personal computers, X-ray machines, supercomputers, files, and personnel across campuses with wireless bridges. Short and long-range wireless telemetry and paging became a force in seismic tracking, environmental, automotive, and retail industries; and wireless banking, stock trading, and credit/debit card transactions became a reality in the late 1990s.

Today, vertical markets are still the most important "home" for business-oriented mobile data exchange; the applications have proven themselves competitively, with hard dollar ROI, improved customer service – even new "models" for profit and revenue generation. (See Chapters 3 & 4 case studies).

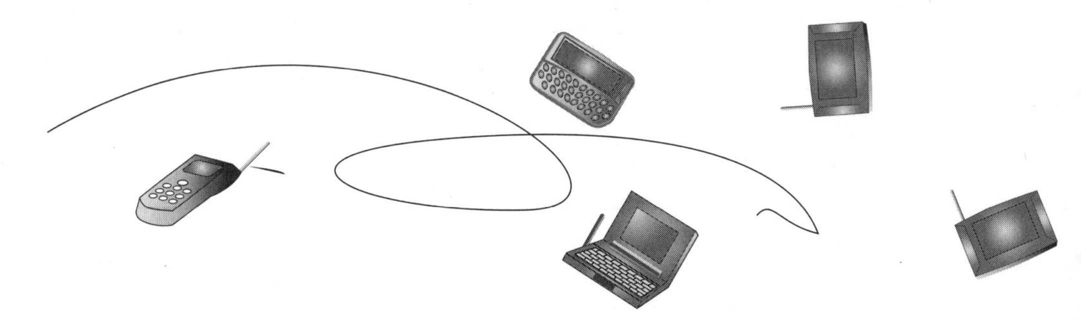

The Growth of "Horizontal" Wireless Business

Beyond verticals, wireless data now tackles a new challenge: the "horizontal" business user, and ultimately, the individual consumer. Sales force automation using wireless devices and links to legacy systems (e.g., CRM or insurance databases for example) became possible with the invention of the Palm VII (a wireless personal digital assistant), the Windows CE-based handheld computers, and, of course, "slimmed down" laptops and PDAs equipped with desktop synchronization devices. Wireless modems, the perfection of "smart phones" capable of sending and receiving data, the ability to download and synchronize messages, calendars, and files from desktop systems, and finally, the invention of wireless Web "browsers," such as *Handheld Device Markup Language* (HDML), now made possible a "white collar" data phenomenon. Business people, many of them executives and managers, wanted an untethered connection to the booming Internet, and, in many instances, the corporate intranet. Systems integrators popped up to "enable" mobile connections behind the firewall. Further, national carriers began to promise the "Mobile Internet," and begin to make it a reality.

In the year 2000, Bank of America Securities predicted that today's 6.6 million worldwide wireless Internet service subscribers would reach 400 million by 2003, according to the Wireless Data forum, *Going Mobile II: The Wireless Web and other Data Solutions*, published by Wireless Data Forum/CTIA, January 2001. Global wireless subscribership was skyrocketing at 30 percent annual growth rates. By 2005, over 500 million users would actively pursue mobile e-commerce (MEC) generating over $200 billion in revenues, predicted Ovum Ltd.

Tomorrow's Wireless

Is the euphoria justified? Probably, although it may take twists and turns of application, commerce, and perspective we do not expect as of yet. The basic drivers for wireless still remain in place: 1) Is it technically do-able? and 2) Is it needed? Providing these conditions are met, wireless data's four key components – devices, networks, applications/content, and gateways – will continue to evolve.

The communication evolution has taken us from simple one-to-one ability (e.g., the early telegraph) to one-to-many over the air (radio broadcasting). We have moved from no security (smoke signals that everyone could see) to highly secured communications with enterprise level encryption and user authentication. We now have hundreds of thousands of conversations simultaneously over the air and the ability, via wireless data, to impart billions of bits of information worldwide. We can roam across country borders while maintaining continuous coverage. There is a weak link, however, in our technology progression. Transmission speeds are struggling to keep pace with content requirements, such as bandwidth-hungry applications like video. This singular factor will be the trigger for wireless data to reach a mainstream audience such as the one mobile voice technology reaches today.

We now see new trends and business drivers:

- *You are how you communicate.* Wireless data is an expression of consumer choice and lifestyle, not simply a tool. The ability to communicate wirelessly is a reflection of our need for mobility, to work flexibly, round the clock, and to take our work with us wherever we go. Wireless therefore represents freedom not only to roam, but also to remain connected, secure, and highly personal ("close to the body"). A wireless communicator equipped with computing capability and Internet access is an "extension" not only of the body, but of the modern mind.

- *Businesses need wireless to increase productivity, customer satisfaction, and competitive edge.* Productivity means time savings, increased revenues, and happier customers: more service calls made, faster order processing, instantaneous package tracking, image transfer, dynamic dispatch, secure access to corporate legacy systems, mobile computing, instantaneous communications, and 'anytime, anywhere' money mobile e-commerce transactions.

- *Wireless initially means higher pain and cost for early adopters, but technology responds rapidly to need.* The pioneering vertical applications of wireless data are primary examples. Initially the services were limited in scope and costly. But with every iteration of hardware, middleware, and applications, RF-driven technology shows its potential for enhancing incumbent technology, even toppling it. Evidence of this is already abundant in the office where wireless local area networks (WLANs) have replaced wired LANs. Wireless LANs are simpler and offer the user the ability to pick up their laptop, move to a conference room for a meeting, and always remain in contact with the corporation's networked services. This technology displacement phenomenon is already happening even though wireless LANs are one-tenth as fast as a standard 100 Mbps Ethernet wired LAN. This displacement is the way a disruptive technology moves from its early market into the mainstream. Users are willing to yield some of the benefits of the mainstream technology for the overwhelming benefits of the disruptive technology.

- *Wireless data creates 'disruptive' business models and new markets.* It reengineers a process, improves one, creates new models for production, services, and revenue generation, offering business and consumers something smaller, cheaper, better, and faster – ultimately disruptive to other business models and production flows. (See our Case Studies for examples.)

- A corollary: *Wireless data creates new channels for revenue generation and cost savings.* Wireless allows new, more effective processes to occur that break geographic barriers, increase record keeping, tie into legacy systems, speed up a process, and make customers and workers happier.

- *Many business processes are dramatically improved by the acquisition and transmission of data in real-time through mobile networks.* Wireless data integrates both data collection and push services, an inflowing and outflowing of vital information. An example: Auto accident insurance adjusters can use wireless terminals and claims processing on the scene to evaluate a claim, acquire centralized actuarial data and records, and also transmit and receive claims, cutting customers a check on site. The competitive advantages of an organizational process like this are enormous; customers flock to business models that serve them better and more directly. Every business has processes that can be improved with wireless data.

- *By its very nature, wireless compresses time and extends space to yield new ways of working.* Real-time, on the fly enterprise messaging is an example – especially in technical or networking companies that service large numbers of customers and need quick staffing solutions in the event of network troubles. An efficient wireless data system can mobilize an entire workforce rapidly, no matter where individuals are located. Further, wireless naturally reduces worker downtime and lengthens the working day – allowing more flexible work schedules. Batch processes converted to real-time processes through wireline connections can now be extended in space and shortened in time to accelerate work and data capture responsibilities.

- *The most important applications for wireless data have not been invented yet.* Right now many voices in the industry project a variety of killer wireless applications such as messaging and e-mail, wireless telemetry, and location services. Given the modest amount of developers who currently have turned their attention to wireless data, it is logical to believe that these applications talk more to our lack of understanding of the power of wireless than to our collective vision. There is no doubt that these applications will prove to be increasingly useful, however, the overwhelming evidence associated with disruptive technologies is that early usage does not predict the larger market the technology may create. We have already witnessed some curious applications that were not predicted. In Japan, NTT DoCoMo's i-Mode service, provides wireless data to more than 21 million Japanese subscribers including an impressive list of applications from 31,000 web sites. It would have been difficult to foresee that downloading cartoons would become one of the hottest applications on the network. Similarly, even at the introduction of short message service (SMS), no one would have forecasted that school students would be using SMS to cheat on tests in classrooms or that billions of messages would be sent each month.

- *Wireless telemetry will become ubiquitous.* We will use sensing devices and wireless telemetry to monitor the living and non-living worlds. Trees, for example, will have sensors that communicate to our home wireless LAN, providing instant information regarding the need for nutrients and watering. Lawn sprinklers will automatically turn when they sense dry ground and have analyzed the weather forecast pushed to them by an Internet agent; a home will be an interconnected universe of wireless sensors, systems, and computing and communications devices. Wireless telemetry will become a regular feature of automobile manufacturing, leasing, and driving. We will monitor the environment, each other (air and space defense), and find wireless data in dozens of other parts of our lives.

- *The wireless user experience will become fantasy-like, if not fantastic.* Picture 'Star Trek' communicators; 'wearable' phones, 'talking data' (speech recognition at natural speech rates), holographic and bubble keyboards, light pens, heads-up displays (attached to goggles and eyeglasses), new screens capable of color and high resolution (HDTV quality or higher). All these devices will become commonplace. Wireless implants, headsets, wrist watches, heart monitors, address books, and other forms of natural interface will change our perception of wireless and make it more transparent to the user. Wireless will get "closer" to the body and will be "worn" like jewelry or eyeglasses.

- *Wireless will become globally connected.* The integration of wireless PANs, WLANs, and WWANs will enable a user to connect at a variety of speeds, at varying costs. Most likely, a form of packet, IP protocol will provide the basis for mobile communications, and a specialized set of enhancements will relieve the chattiness of wireless conversations. PANs will become commonplace, enabled by the Bluetooth short-range RF standard; and a myriad of application providers who write applications on top of Bluetooth will automate our lives. Printing while mobile, for example, will no longer be a problem.

These are the emerging patterns of change we are seeing right now. Now read on to find out how pioneer adopters untether their enterprises, reengineering business and communications to get a competitive edge. Remember, these applications are only a sampler of what is being done today across a wide a range of application types. Even then, creativity will be the limiting factor to how wireless data can be used in the enterprise for profit and competitive advantage.

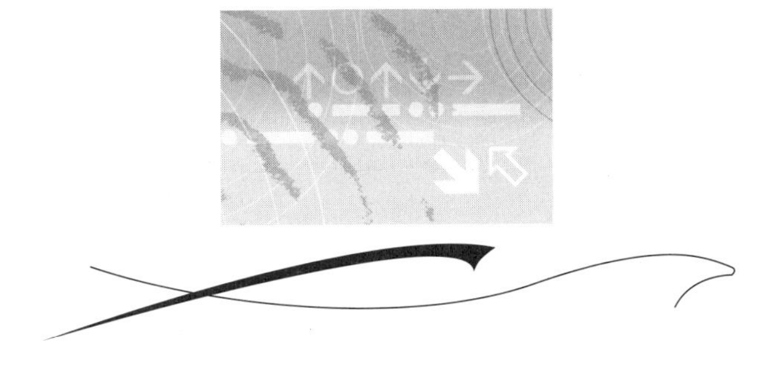

MOTOROLA
StarTAC

3 UNTETHERED ENTERPRISE: NO PAIN, NO GAIN

"Beam Me Up, Scottie"

THE FIRST ENTERPRISE ADOPTERS FOR WIRELESS data had a keen sense of mission and a lack of pragmatic experience from which to devise a workable solution. Wireless data was still on the fantastic side — a "Beam Me Up Scottie" notion of tiny communicators and invisible frequencies carrying voice and data in conversations between humans, devices, and network servers. If only someone could figure out what wireless networks ought to do, and how exactly they should perform, an enterprise could enter a new unwired life and actually make money. In reality, though, the early pioneers had little to work with. FedEx, for example, built a proprietary wireless data radio network in the 1980s because the company needed a more customer friendly way to track packages on the move – and no other outside firm at the time provided a suitable answer.

Sears also decided to revamp its field service operation in the late 1980s by using wireless technology. The company realized a service technician spent valuable time seeking out a dial tone (e.g. pay phones and customer phones) to check stock, receive dispatch assignments, and order parts. On average, technicians spent 30 minutes a day with these calls, and efficient dispatching was hit and miss because of the number of customers who would cancel appointments after a technician was already en route. Clearly, the company needed a real-time system of mobile communications and data interchange. Nonetheless, it took several years of research, experimentation, seeking out partners, conducting trials, and modifying network technologies before Sears Product Repair Services implemented a practical wireless data application. Even after the first deployment to 7,000 technicians in 1992, the company added new network providers (supplementing ARDIS with RAM Mobile Data and NORCOM, a satellite network provider). Within a few years, devices had changed from IBM ruggedized terminals to Itronix ruggedized PCs; and a new middleware provider, Nettech (now Broadbeam Corporation) wrote interfaces that moved the network from ARDIS and DOS operating system support to Windows and support for multiple networks, including satellite. Sears' concept for mobile networking evolved rapidly, too, as did expectations and demands for better network coverage and service efficiency. By 1998, the Sears field service network reached 12,500 technicians. It offered 100 percent employee field coverage through its terrestrial and satellite networks, enabling technicians to eliminate paper work and reduce phone calls for parts and information by 50 percent.

Bottom line? Sears Product Repair Services' wireless data application increased the number of service calls per technician, transforming operations from loss leader to profit-center. Yet the "journey" was not easy; it had no precedent. Implementing a solution was not without periods of wandering through the woods, stumbling down blind alleys, and wondering if there was life after analog cellular.

Good Questions are the Start of a Solution

The beginning of the "journey" to an untethered enterprise always starts with questions about the purpose and form of the application.

Here is an essential list:

- *Can wireless technology improve the efficiency or effectiveness of our equipment or personnel?* If so, how?

- *Can wireless data help us build a new type of business or profit center, or to automate/improve existing processes?*

- *Can we use wireless to build the most important intangible – customer satisfaction?* How can mobile data (or mobile access to enterprise data) help retain our customers, reduce churn, and squeeze more out of the business process?

- *What is the likely return on investment (ROI) for a wireless data network?* Can it be demonstrated as tangible ROI, or as business "intangibles," such as improved customer service and employee satisfaction?

- *How can wireless technology help us win new customers and differentiate ourselves and our offering, giving us a competitive edge?*

Those are the positives. Now here are some "can we do something else?" questions that could, at the outset, help you find options other than wireless data to solve the business problem:

- *Could I use a wired network to accomplish the same objectives as a wireless network?* If so, a wireline network is almost invariably a less expensive alternative with more options for network infrastructure, data sharing, and access. However, there are exceptions to this rule. In some cases, a wireless network can prove to be *less costly* over the entire life cycle of an application – when such factors as maintenance of a wired physical plant, moves, adds, changes, plus labor costs, and software upgrades are factored into the analysis.

How rich is my content? If the data required for "shipping" wirelessly to a mobile user is extremely rich, requiring gobs of bandwidth for transmission (e.g., streaming video or complex graphics), a mobile wireless network will not be "do-able" or satisfactory – at least not now – barring breakthroughs in signal encoding, compression, or the availability of 2.5 or 3G networks. (Some of these advanced networks will be tested and implemented this year and next – and will support high bandwidth mobile applications, such as very large file transfer and streaming video.)

- *How much bandwidth does my application really require?* Today, some specialized mobile wireless networks (e.g., Metricom) operating in parts of the country can support business quality video and data at up to Integrated Services Digital Network (ISDN) speeds (128 Kbps). But specialized wireless networks available nationwide do not exceed 19.2 Kbps (examples: CDPD and ARDIS) – the speed of a "slow" wireline dial-up modem. Network congestion and over-the-air errors causing resends of data reduce this maximum transmission speed.

- *Can my legacy application be adapted for wireless?* Sometimes yes, sometimes no. Back-end integration with new mobile wireless systems and legacy applications generally requires system integration skills. Some wireless application platforms simplify this problem by providing integration points with existing information technology constructs, such as databases, enterprise application integration (EAI) software, message queuing software (e.g., IBM's MQ Series), e-mail systems, and specific applications (e.g., Siebel Systems or SAP). Another technique for reusing an existing application is to use a transcoding server, which renders the application's user screens for each type of mobile device. This technique works best if the source application is written in eXtensible Markup Language (XML).

- *What will the user experience be like?* If you've identified the application and the available devices and networks to run it on, consider what the experience of wireless users will be like. If the user experience is poor (through trials or investigating comparable user experiences in already existing networks), chances are the user will become dissatisfied and disinterested, and you will have wasted a precious investment in time and money. User experiences are dictated by many factors – the device, the application, its ease of use, the availability of various touch screen and keyboard interfaces, the amount of input required by the user, the quality of network coverage, quality of transmission, among others. Consequently, before launching the application, consider looking in depth at design and application trade-offs. If the user is unhappy, chances are the wireless solution will not be accepted.

- *Are all the elements of my solution known?* In the case of many new wireless applications, the overall number of pieces that must be assembled are not well known. Therefore, the risk of creating a new solution must be weighed against the potential rewards and alternative options available. For example, when assembling a wired LAN today, corporations can avail themselves literally of hundreds of thousands of network-certified engineers. In the case of wireless, however, very few people understand all the complex pieces – the networks, the end user devices, the application software and wireless middleware. Therefore a business team must consider another question:

- *Do I have a staff to design, build and manage a wireless business application?* If not, how do I find the talent to execute and create one? In other words, will a wireless data solution be managed in-house or by outside help? A large Fortune 500 or 1000 company might have an IT staff and the interest to write applications and deploy new networks. A small to medium-size business is likely not to have that expertise, especially in the case of groundbreaking wireless technology. Therefore, managers must figure out who will take the lead on a proposed groundbreaking project – and how tasks will be allocated and integrated. In the case of many wireless solutions, finding a common denominator – and an able systems integrator – is a key to success.

Because the "journey" to a solution is a process, it's necessary to imagine, plan, provision, retool, and modify concepts expectations as you proceed.

In general, wireless middleware providers often touch every aspect of a wireless data project and may be in the best position to work creatively with carriers, enterprise customers, and the application development team. Wireless application vendors, however, are often willing to act as integrators and consultants and can be an excellent source of information. Most vendors will take the time to educate prospective customers on available options – but within their realm of expertise.

The "Five W's"(and one "H") of Wireless Data

If we think of wireless as a delivery mechanism, then, what really matters is the data or content required – *who* will receive it and *where, what* and in *which* form) it must be, *when* it has to get there, *how* it will get there, and *why* (the business benefit). It may sound like a familiar mnemonic from journalism, but in any requirements analysis, a first step is to understand the "five w's" of network before jumping to the "how." In an analysis, the basic work comes in determining: 1) which information do "we" (the corporation, IT department) need to put into whose hands (e.g., employee, customer, supplier) under what circumstances; and 2) which information do we need to put into our hands to make the wireless application valuable to our business? In short, managers must look at two-way information flow to design a system that produces a tangible ROI and a set of powerful business benefits. The end result of an "on the mark" wireless data application is that the business becomes more responsive to the customer, enterprise, end user, and supplier, delivering better, more accurate information and service. Moreover, IT departments can minimize risk by conducting small trials of a wireless implementation with in-house staff and users before announcing and launching commercial service.

Three "Dead Giveaways" for a Wireless Data Solution

When all of the initial questions, caveats, and ruminations are said and done, *three business factors* will almost invariably drive a wireless data project given the appropriate resources and technology expertise:

When employees or users are highly mobile. Mobility is primordial. If users/employees are roaming and require mobile data communications on the go to automate a process or work more effectively, wireless data is an attractive technology. Alternately:

When the customer needs to reach you quickly and get a substantive response. At one time, experts thought that wireless was principally a boon to vertical industry. It offered immediate real-time, on-the-fly mobility solutions for field service dispatch, telemetry, inventory tracking, and point of sale. Then wireless became a means of selling and delivering "e-mail" and messaging. Today and tomorrow, a killer application is likely to be touching and communicating with customers. If wireless makes it easier for the customer to reach you and your suppliers, it has an immediate business advantage and justifies an investment of appropriate scale. Wireless data can be used as a means of bypassing the classic phone-tethered service rep. It also enables customers to tap a few buttons on a mobile PDA to get goods and services, order dinner, get theatre tickets, or stay in contact with people.

When corporate data must be delivered on a real-time, 24 X 7, 'anytime, anywhere' basis to employees, suppliers, customers, or contacts. Ubiquitous, real-time data delivery on a mobile basis that follows security protocols is not only possible with wireless networks, but may be the only way to give enterprise the competitive edge in the markets that they serve.

Remember: Round-the-clock availability, mobility, and customer responsiveness are three big factors that spell success for organizations implementing wireless data networks. Those factors alone can drive a solution even when technology constraints appear to narrow options or point to options as yet unknown. Because the "journey" to a solution is a process, it's necessary to imagine, plan, provision, retool, and modify concepts and expectations as you proceed. We will show how various pioneers have accomplished these steps in the next section and in Chapter 4.

Pioneers: Fidelity Investments Go Wireless

Fidelity Investments began an enterprise messaging program in 1997 to alert technicians and programmers via a paging network of job "abends" (abnormal ends to programs which translates to outages in production job streams) requiring immediate attention. The wireless application was written in-house. "It was not very sophisticated, but we had embedded a code directly into production processes to alert technicians wirelessly of an outage," explains Joseph Ferra, senior vice president and chief wireless officer, Fidelity Investments. Today, that internal messaging application has transformed into a nationwide wireless trading network for retail customers, called "Fidelity Anywhere." As of April 2001, Fidelity had serviced more than 92,000 customers, with thousands more non-customers using the service for market quotes and watch lists.

Early Stages: Assessing the Customer

Running on Skytel and other paging networks, the first application alerted technicians at various regional sites of issues at the Fidelity National Data Center. This was a starting point, and Ferra and other managers hit on the idea of conducting a large-scale survey of very active investors (those who traded 36 times a year or more) to assess their needs for real-time, mobile information. "Thirty nine percent (39%) indicated they had missed vital trading information. Had it been available to them, they would have acted on it," Ferra said. The response prompted Ferra to look further. "At the time, no one was offering direct wireless links to personal investment accounts," he said. "But by October 1998, we had already delivered a simple way for individuals to use wireless to get account balances, cash positions, and valuations, and we were clearly the first to offer personalized delivery to a handheld device." Immediately following the first offering, however, customers began to inquire about on-line, real-time wireless trading. "But we weren't ready to launch it in 1998," Ferra said. "The technology out there wasn't sufficient to provide simplicity and the encryption we needed to ensure security for the accounts."

First Deployments: Going for Simplicity, Security

How did Fidelity Investments evolve a solution? "At the time, in 1998, there were no Palm VIIs or Palm Vs with Minstrel Modems, no Omnisky portals, no Web-enabled phones or other PDAs with wireless modem attachments," Ferra recalled. Although he saw a proliferation of handheld devices on the marketplace through 1998 and 1999, "it was difficult to offer a Web-enabled service because the networks didn't have nationwide coverage, and their protocols may not have been in line with roaming partners." In some cases, devices were actually ahead of the available networks (e.g., the Motorola I1000+ phone, which included walkie-talkie capability, wireless Web capability, and paging).

The company opted for simplicity: two-way paging devices from RIM and elliptical curve cryptography (ECC) encryption from Certicom. "We knew we needed not just the right coverage and data rate, but confidentiality – therefore encryption of data entered the picture," he said. Most encryption schemes lengthened the wireless message, making it ungainly, resulting in "tremendous network degradation and data latency," he said. However, after considerable canvassing, RFIs and RFPs, Ferra's department commissioned the wireless security provider Certicom to embed cryptography code into the system. "ECC passes a key from network server to handheld device so only that device will know the encryption scheme," he said. Fidelity collaborated with Certicom to implement this form of elliptic curve cryptography efficiently for paging, and a layer of encryption was added to the BellSouth Wireless Data network (now Cingular Interactive), making it much more secure. Ferra's staff also opted for the RIM 950 Interactive pager for the customer device, which was able to accommodate dedicated two-way transactions in an encrypted form.

Fidelity maintained an exclusive contract on the BellSouth Wireless Data network for calendar year 1999. Fidelity chose not to handle the wireless device or billing aspects of the service, instead partnering with BellSouth Wireless Data, which managed the subscriptions and hardware delivery to customers wanting Fidelity's wireless service. "We put the application out there and set up a system so that our customers could contact BellSouth to sign up for service," Ferra said. BellSouth handled order fulfillment and would ship the RIM 950 Interactive Pager and start-up kit directly, while Fidelity spread the word of its wireless services through customer service representatives, information on its web site (www.Fidelity.com), and print and broadcast advertisements.

Ferra says Fidelity was the first to offer the personalized financial information service to retail investors in October 1998. Other companies such as Reuters and Discover Brokerage were also early starters in wireless financial services. By January 1999, wireless trading began. "We first offered this service to our most active traders, but by February of 1999 we had expanded the offer to anyone with $100,000 or more with us in the firm, and by third quarter of 1999, we had broadened the offer to everyone of our customers who wanted it."

Fidelity also struck a unique agreement with Palm Computing (at the time 3Com), to embed the Fidelity application into every Palm VII in the United States. Fidelity offered personalized account information, trading, and the ability to have non-Fidelity customers, but Palm users, access Fidelity for market information and watch lists.

A Good Combination: In-house Talent, Strategic Partnerships

The IT staff at Fidelity created and executed the software for this wireless network – an ability augmented in part by a large IT staff, the availability of wireless experts in-house, and corporate experience with a successful wireless enterprise messaging application. Fidelity's only partners on the initial project were BellSouth Wireless Data and Certicom. "We had a lot of discussions with RIM, which actually built the handheld device," Ferra said. "But one of the things that was really important – something we observed and learned – was the element called over-the-air programming (OTAP). We were asking ourselves how we could make continuous upgrades to the wireless data application. After the initial installation, this got tricky, especially with folks continuously getting on networks and adding new devices. But RIM was very helpful in allowing us to do OTAP."

Ferra's staff realized it wasn't practical to ask customers to bring their devices into one of Fidelity's investor centers every time new wireless trading software was released. So upgrades were accomplished using OTAP; Fidelity customers automatically received a wireless notification of an impending upgrade (e.g., new features that permit the customer to establish stop loss limits or access net benefits from a 401K retirement account).

By spring 2001, Fidelity counted 92,000 subscribers to its wireless service. "We now have it on every major wireless device and network," Ferra explained. "When we launched the service in January 1999 with BellSouth, it was nationwide and available to anyone in the continental U.S. Palm originally marketed its Palm VII on May 24, 1999 in the New York metro area. However, if someone bought a device in New York, and brought it home to, for example, Illinois or Massachusetts, he or she still could use the device because it operated over the BellSouth network. Palm made the device available nationwide on October 4, 1999."

Every Palm VII device manufactured today has the Fidelity wireless application built in, and Fidelity has also made it available on Sprint phones through the Sprint CDMA network, the CDPD Omnisky wireless portal, the Verizon CDMA network, Nextel, and OnStar's virtual advisor service in automobiles. Fidelity customers can get a full array of services and transactions: trading and personal account balances, quotes, market information, percentage gainers, losers, most active, order status, and net benefits. "'Net Benefits' refers to the Fidelity service that allows employer-sponsored retirement plan participants online access to their retirement accounts from 401K," Ferra said. Even those with Fidelity retirement accounts who don't have retail brokerage accounts can now subscribe to the wireless service. Expanding the venue has significantly increased customer satisfaction, Ferra says.

Spiraling Expectations and Benefits

"If we look at our wireless customers in comparison with the average retail brokerage customer, the wireless customer is a desired customer, even though that's a small percentage overall," Ferra says. "The asset base is higher, the wireless customers' number of transactions is higher, and their use of various (and multiple) channels is much greater (e.g., web, phone, speech recognition, wireless service, visiting various Fidelity branches). So we're looking at someone very affluent, relatively active, and multichannel." Why is this important? Fidelity believes the strategy is essential not just for customer retention, but for new acquisitions. "We are averaging 400 new account openings per month through the Fidelity Anywhere channel alone," he reports. "Although it's difficult to capture exact ROI at this early stage, we are identifying quality customers and adding new ones."

Wireless is tracking in a similar pattern to voice response in the '80s. At first, customers didn't use it much. Now voice response has become not only a standard service feature, but has improved customer service, Ferra says. The new tools create spiraling expectations. With financial competitors also offering wireless, Fidelity must evolve newer, more robust services. "Wireless customers now have the ability to gain real-time access to personal data and retirement account information that is not location dependent," Ferra says. "So we've changed our business model for the wireless service – not just to retain but to acquire entirely new customers, and we've had good success."

Bottom line: Fidelity's Anywhere wireless data is a perfect example of how a company can woo customers and gain competitive edge by offering 24 X 7 *transactional power – the ability to access and move on personalized financial information.* While it took some work to get there, Fidelity established a first mover's advantage, which is hard to displace. The company will continue to lead the financial services market with disruptive technologies and applications.

Pioneers: Unwiring BellSouth Telecommunications– "May the Field Force Be With You"

Fifteen thousand field service technicians are part of BellSouth Telecommunications, the support organization for wireline phone network operations in nine states in the southeastern United States. Now, thanks to a massive wireless data implementation, the field force is able to access up-to-the minute customer service and network status information, speeding transaction times and saving on average 45 minutes per day per technician. The solution uses Cingular Interactive's Mobitex network, Broadbeam's wireless messaging server, ExpressQ, Itronix X-C6250 ruggedized laptops (see Figure 3.2), and a custom-built field service dispatch application created by BellSouth using services from Andersen Consulting (now Accenture). It has taken seven full years to evolve the success it enjoys today – being a pioneer has its rewards.

In 1994, BellSouth was beginning to look at a mobile computing platform for service technicians because the company's field workers were having difficulty finding dial tone to link with dispatch centers and parts and inventory. "We were probably at a better vantage point than other service companies because at least we had the familiarity with where a phone was in a particular subdivision or industrial park," noted Gary Dennis, general manager of BellSouth Telecommunications' network operations solutions group. "But the technicians often had to use customers' phones to obtain trouble tickets and service orders, so we looked at three options at the time, and the most logical starting point was, 'Why don't we use a cellular network to transmit circuit-switched data? "*That* was a blind alley."

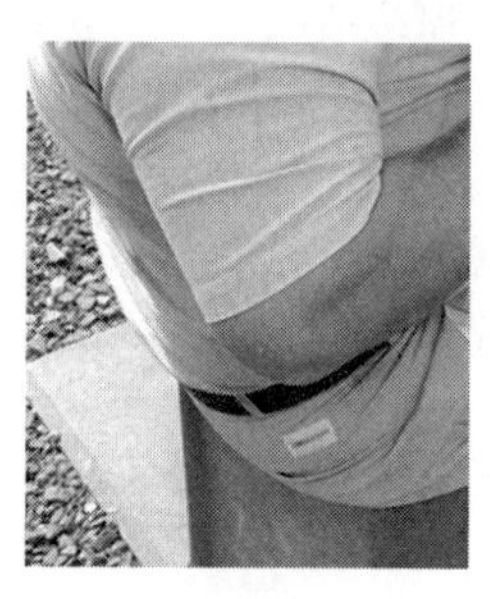

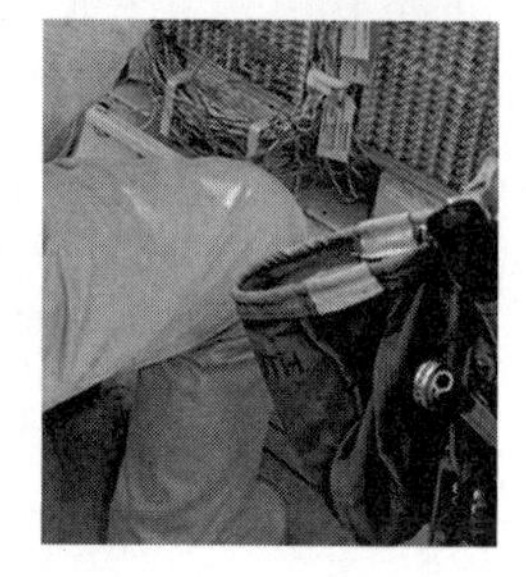

Figure 3.2 BellSouth Field Technician Connects Using Wireless to Expedite Trouble Ticket Resolution

At first, Dennis and his staff tested a cellular solution with 20 technicians who carried laptops equipped with wireless modems (i.e., PCMCIA cards). However, they found that connections were often lost due to coverage problems and dropped calls associated with circuit-switching. "With basic circuit-switched technology, we found you could drop a call just by running into a bad cellular pocket. By the time you type in the passwords to get access to the computer (and data) in the data centers, a minute could go by. You'd start a transaction, then get a dropped call and have to start all over, so we saw the difficulty not only in the time lost and annoyance with dropped calls, but more important, the cost. The worst of it was the cell phone bill was several hundred dollars per month per technician, and that was just for the cell phone calls to these back office computers to conduct tests," he said. A technician could be on a call for five to eight minutes just to get test results; and with separate dial ups to the dispatch system to get service requests, access e-mail, and dial up the computer to do time reports. Dennis found the cellular solution was basically impractical.

Trying Other Approaches: Packet Data Proves Economical

The Network Operations Solutions group tried other approaches, considering CDPD technology at the time. However, the CDPD option was quickly eliminated because BellSouth Mobility had not deployed CDPD in the region at that time. "We also looked at the BellSouth Wireless Data (now Cingular Interactive) Mobitex network, but we were nervous about the 8 Kbps throughput of the network," Dennis recalled. It was, however, a specialized packet network just for data, and "that was a breakthrough for us, because it gave us the data communications we needed." The dropped call problem completely went away after deployment. "That's because with packet technology, if you lose a connection, as soon as you go back into coverage, the transaction is completed," he explained. This happens because the individual packets of wireless data in a transaction are "addressed" and routed individually; the transaction does not require that the network "reserve" a dedicated circuit to complete the connection. Instead, individual data packets are sent, finding their way to the destination via different routes, so a temporary lost connection does not require a repeat login or transaction from the starting point. Best of all, the solution proved fairly economical. "We designed a client/server approach (the client was the technician's laptop; the server was a Sun Microstation back in the data center)," Dennis said. (See Figure 3.3)"Our phone bills with cellular had been averaging over $200 a technician – way out of range for a business case that was practical for us. However, it looked like we could limit the transactions if we did it with a packetized arrangement, and used a 'thick client' (a laptop equipped with a lot of applications software, memory and processing power). So we went with full-scale laptops instead of a handheld device like a Windows CE or Palm. We wanted, in other words, to minimize the data transaction between client and server, so that costs were in line."

Figure 3.3 BellSouth Field Technician Connects to Main Office

Shortening Transaction Times

The packet switched strategy worked: Dennis was able to test a network to the tune of $30-$3500 a month, on average, per technician, for Mobitex coverage. "We were actually under budget the first year on communications costs, because even with technicians who weren't quite familiar with the system – and the hump of the learning curve – we beat our budget expectations. Because we had put a lot of software on the laptop, that meant we could limit the data between the laptop and the final destination in the operations systems in a remote data center," Dennis explains. Example: data templates were already loaded on the laptop so that all the wireless connection had to do was "populate" existing technician forms with new data.

"In effect, we're only sending pieces of a screen that need to be updated, and that kept communications and bandwidth requirements low, so the 8 Kbps concern went away too," he said. "Our data fits very nicely with the lower data rates, and keeps the cost down. That has really been a non-issue. So the third part of the equation – coverage – ended up on target with Cingular Interactive covering 90% of the day. In other words, nine out of 10 times technicians could execute a transaction when needed. "The carrier took the challenge very seriously, so when we started to roll out in the fall of '98, completing most of the roll out in '99, we held Cingular to the standard," Dennis recalled. "Cingular embarked on a major construction project, which took them about a year and a half. We knew where all the work centers were located and helped the wireless data people design where the additional cell sites needed to be; and then we hired a separate contracting company to check on the coverage."

"Cingular Interactive delivered as promised," he said. "They stepped up to the plate, and we've had a very successful rollout as far as coverage goes thus far."

Drivers for the Solution

Several elements came together at once to drive the field force solution. Around the time Dennis began discussions with Cingular Interactive, the release of the highly successful RIM 950 Interactive dramatically increased pager subscribership to the Mobitex network, giving Cingular Interactive the boost to build out its network and create a business case for cell site additions. With 15,000 technicians averaging $30 a month in wireless network charges, the relationship between BellSouth and BSWD was a win-win. Internally, the network was called "TechNet." (Later the software application would become TechPlus, which was sold to Telcordia. Telcordia now markets the capability under the product name, TechAccess.)

Originally, BellSouth Telecommunications had used middleware developed by a small company that went out of business. But following the first pilot of '97, BellSouth opted for Broadbeam Corporation to provide wireless middleware expertise. "We built software for telephone company applications, such as a service order system, but decided we needed Broadbeam to develop the wireless interface between the field force application and protocol servers (where the middleware resides at the data center)," Dennis said. BellSouth Telecommunications wanted an off-the-shelf solution offering advanced wireless messaging features, including store-and-forward capabilities that make it possible to send data automatically once coverage is reestablished, as well as support for multiple networks. Assisted by Zamba Inc., an independent contractor working with Andersen Consulting, which consulted on the project, BellSouth chose Broadbeam's ExpressQ™ mobile messaging middleware. The new middleware solution provided a new level of communications security.

"In '98, we had another major change: a new middleware provider; and a new wireless modem provider," Dennis recalled. "Initially we had bought Motorola modems with our Itronix laptop purchases, but Motorola decided to get out of that business. So in '98, after having done the pilot and bought five or six hundred Motorola modems for the laptops, we changed over to RIM and its wireless modem capability. We bought enough RIM technology for 15,000 technicians in 1999."

The entire TechPlus implementation cost $163 million to provide the technicians with both wireline and wireless connectivity to their home bases. Initial capabilities for the network included on-the-fly dispatch, receiving and closing out trouble tickets in real-time, and completing work orders. Technicians also gained immediate mobile access to existing BellSouth computer systems for network testing, work assignments, maintenance, and customer information. Today, as many as 58,000 wireless transactions per hour are piped through the system, which is entirely IP-based in the back end.

Productivity: A Confounding Process

BellSouth's fleet of more than 15,000 customer service vans has been equipped with battery charging stations for the Itronix notebook computers, as well as HIGH-gain antennas and full-sized computer printers. The network utilizes multiple Windows NT Servers running the ExpressQ middleware, providing redundancy and the ability to scale. "In all," Dennis said, "our transaction times are faster and we have reduced network transaction time an average of 48 minutes a day per technician, which has been scientifically tracked," he said. "We've reduced the number of log-ins and that should translate into productivity gains." However, BellSouth Telecommunications, like many other phone companies in transition, has experienced huge technician turnover – as many as 8,000 of the 15,000 technicians on the job are brand new since the wireless data network went live. This has confounded research on productivity; since technicians must learn on-the-job skills before productivity increases are fairly assessed.

"We have some real productivity improvement with technicians with over three or more years experience," Dennis said. "The middleware, in particular, has proven to us that it works and can keep a computer accurately and reliably connected to computers in the field. So now we'll have laptop PCs connected to Web sites, and our next generation of TechPlus will rely more on Web-based applications to the back office. We'll use browsers on our laptops, and give technicians access to Web sites for new information and training programs they can download." The middleware will be modified to enable interaction with multiple Web sites. "I assume that when speed becomes a problem, we'll work on upgrades to the network. But right now we're just getting our feet wet. We're designing for interactions over the web sites that are still working well wirelessly with the available data rate (8 Kbps link) with less than 10 seconds delays. In sum, "the TechPlus system has given us an efficient competitive edge. This should translate into a way to stay on top of the competition."

Your Enterprise: Where to Go from Here?

In a September 29, 2000 Equity Research report, entitled "Out of Thin Air" (J.P. Morgan Securities Inc. Equity Research), author/analyst Paul Coster argued that "enterprise-class solutions are in a sweet spot of the emerging wireless industry sector. As value-chain participants, [these applications] should benefit from both the build-out of wireless infrastructure, software and services (the journey), and from running the infrastructure for the end-state B2C (business to consumer) business models. We also believe that corporate demand for enterprise solutions will take off in the next 18 months," especially where there is a definitive and measurable ROI.

As already outlined, the key to virtually every enterprise-class solution is that four major elements must come together: device, network, application (content), and gateway services (e.g., middleware). Because enterprise wireless solutions are complex, they must "choose" between, or integrate many elements. These include multiple network protocols, multiple device environments (e.g., PDAs, laptops, palmtops, and smartphones), and diverse corporate data systems (e.g., UNIX, Windows NT, Linux, POP3, ODBC databases, SOAP, J2EE, Lotus Domino, Microsoft Exchange, HTML, legacy systems, et cetera). The solution must integrate software to handle such variable issues as customer care, provisioning, network operations, airlink optimization, security, carrier agreements, and maintenance.

There are significant differences in the definitions of networks and requirements when approaching solutions from an enterprise to worker angle vs. an m-commerce, consumer angle. Example: employers may control the device and networks that an employee uses, but have no control over how a consumer may access the firm's m-commerce systems. A business to work application can be highly engineered and include customized software on a chosen RF device, whereas an m-commerce system or service will most likely be device-agnostic. Increasingly, enterprise class solutions are driving toward device and network agnosticism, with many complex considerations taken into the requirements analysis.

Bottom line: IT managers must make unique, in some cases, never-before -tried choices about networks, devices, content, and gateway options. Much will depend on the level of service/performance and security required – plus the needs to maintain enterprise control and an adequate level of user satisfaction and efficiency. In many cases, timing is everything: a solution cannot be devised until numerous technologies come into alignment in a favorable way. As wireless data technology evolves, more combinations of winning elements will in fact be demonstrated. But without a certain amount of uncertainty and pain, there is no gain for anyone – not pioneer adopters, and certainly not for first-time users and their managers. Wireless data, in short, is no longer a fantasy, yet in many ways it continues to be an experiment.

In the next chapter we will examine five leading-edge case studies – all enterprise wireless data solutions, all successful – showing how these critical elements came together realistically to produce tangible gains for the enterprise.

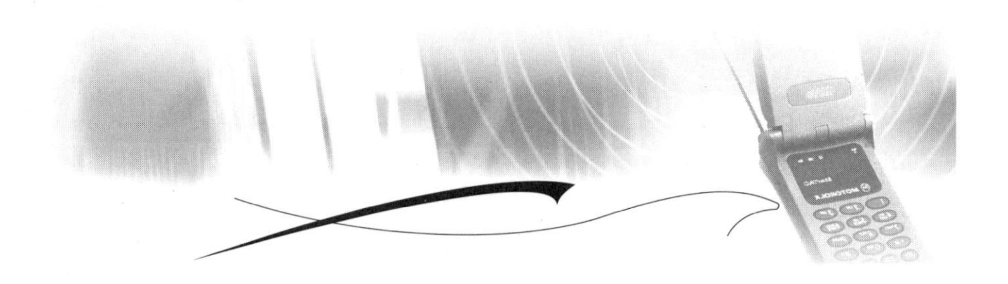

355 67797 3597678 0009454365768234

4 AT THE LEADING EDGE

FedEx: A Revolution in Wireless Package Tracking

Business Challenge: Better Dispatching

ONE OF THE EARLIEST INNOVATORS IN WIRELESS data usage was the "When it Absolutely Positively Has to be There Overnight" company — Federal Express Corporation

(FedEx). The company had used a traditional voice dispatch system since its operations went "live" in 1973. But as business soared with 30 percent annual growth rates per year, the system became overwhelmed because of the volume of voice calls between dispatch operators and drivers. "We were on shared systems, and growing so fast that they quickly became unmanageable," said Winn Stephenson, a senior vice president of information technology for FedEx Services. The company sought another approach: wireless data to drive greater efficiencies in data capture, dissemination and responsiveness to customers. "Necessity being the mother of invention, we decided to go ahead and get our own dedicated channels," Stephenson said.

Solution: En Route Tracking of All Package Flows

At first FedEx did not envision a system for tracking packages at numerous turn-around points on the route. The focus was to assemble the elements of a pure data-driven mobile courier dispatch system using private radio channels. However, by 1986, the strategic direction expanded to enable "tracking for all package flows," Stephenson explained. "The radio system then ended up being a front end data collection system. The couriers would pick up a package and upload that information into the database – all done wirelessly by 1986," he said. "We modified all the components to accept end-to-end package tracking, including making changes to the *Digitally Assisted Dispatch System* (DADS) terminals and writing new software to move the information upstream from the SuperTrackers® (courier scanners), to the terminals in the trucks. We then modified the applications so they could transmit the package information back to the databases."

By 1986, FedEx had written its own wireless middleware for the project and created a robust internal wireless engineering group to manage 500 repeater sites and 35,000 terminals, a nationwide private radio/dispatch network that was second to none.

How The Application Became Disruptive: Earliest Stages

The company assembled its first system in the early 1980's by locating Mobile Data International (MDI, part of Motorola), a Vancouver, B.C. company. MDI provided a data terminal and transmission system for FedEx trucks; its interfaces were similar to the keyboards and CRT screens used in early 80's police cars. Transmissions ran at 4800 baud using MDI's MMP-30 proprietary protocol for FedEx's private radio system. "We wrote our own application that the dispatchers would use to manage the mobile fleet," Stephenson said. The advent of the digital system, DADS, in 1980 enabled FedEx to manage two to three times the number of carriers/truckers – more than 125 couriers per channel (with one dispatcher handling each channel). "Under that system, we'd get 2 1/2 times the effective use of each DADS channel. The terminal would hold the information, so while the courier would go out for pick up or deliveries, the new/updated work schedule arrived to give him the next stop," Stephenson explained. "The data was always received perfectly and correctly; there were no miscues on addresses; and the couriers loved it because they didn't have to call in all the time." The system was so efficient that often while a FedEx agent was on the phone with a customer processing his information, a courier would be dispatched at nearly the same time to pick up the package. "A courier would often show up at the front door while the phone conversation was still going on. Talk about customer service!" Stephenson said.

Advanced Stages: Wireless Package Tracking at 'Turn-Around' Points

Through experience, it became clear to FedEx that the technology could actually change the model for customer-shipper interaction by enabling customers to keep close tabs on packages as they moved along the route. By 1986 FedEx changed its wireless objectives to encompass tracking of all packages throughout the system. Private radio coverage had been extended throughout major metro regions and to about 90 percent of all FedEx territories nationwide. The company maintained its own middleware and had modified its system to enable dispatch and package information to move from scanners back and forth into databases, allowing a minimum of nine different data checkpoints throughout a package delivery cycle.

"After 1986, we made incremental improvements – put in higher speed technology and modified components and terminals," Stephenson reported. "In 1996 we implemented a proprietary network, running at speeds of 19.2 Kbps (see Figure 4.1 below for System Diagram), which was similar to a CDPD network in being 'always on' and available. FedEx used it not only to allow instantaneous communication between its dispatchers and courier force, but also to retrieve information about packages scanned for customer tracking inquiries. This advantage became "disruptive" for the entire "overnight" package shipping industry and set an entirely new standard for effectiveness and customer satisfaction. "Customers want to know about their package and we can give them that information every time the package changes conveyance," Stephenson explains. "The average package was scanned 12 times in stations, trucks and hubs; so we are on top of that; and all that information was collected in our databases and available to our customers. The impact is tremendous – ideal for customer service and for our internal operational needs." FedEx's wireless tracking proved to be its biggest advantage over competitors. The wireless network delivered a customer service coup, enabling the company not only to keep customers informed about their package, but also to provide them real-time information.

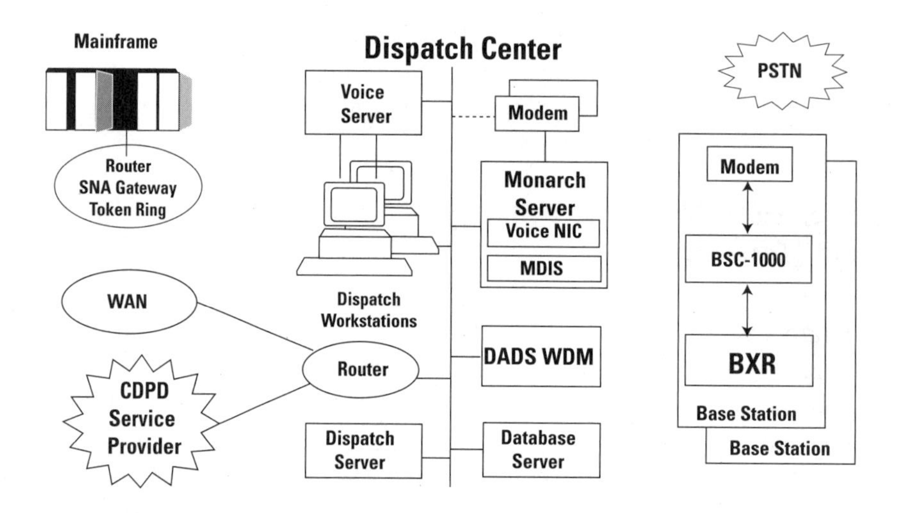

Figure 4.1 FedEx Monarch System Block Diagram

Networking Set-Up

The original DADS systems were developed and implemented between July 1980 and August 1982. At that time the FedEx DADS system covered about 80 percent of the company's markets. The first major upgrade to the RF portion of the network came in 1986 with the move from analog to a fully digital network. The upgrade improved throughput by a factor of four, gave FedEx the ability to transmit voice and data simultaneously, and provided roaming features between the company's repeater cells.

As of August 2001, the FedEx DADS network includes 350 dispatch centers and over 750 base station and repeater systems nationwide. At that time, over 40,000 courier vehicles and nearly 10,000 foot couriers communicated over the DADS network nationally. The system is a cellular packet data network transmitting 50,000 megabytes each month. By August 2001, FedEx covered over 95 percent of its markets with DADS.

Configuration

The network has a CDPD-like (Cellular Digital Packet Data) protocol called FMP (FedEx Mobile Protocol). Like CDPD, this corporate network operates at 19.2 Kbps. Unlike CDPD, the network handles both voice and data simultaneously, allowing some couriers to use voice while others are uploading package scan data at the same time, over the same channel.

Referring to the system diagram in Figure 4.2, the DADS system contains three major subsystems: 1) the centralized computer customer application (COSMOS) with associated mainframe and wide area network connectivity to the 2) Dispatch Center, which contains the voice and data servers for the dispatch workstation along with routers and servers for the 3) radio network, which includes base station controllers, transceivers, antennae, and subscriber units.

Lessons Learned Along the Way

The early implementation of DADS was challenging, since this capability did not exist at that time. FedEx pioneered mobile data by modifying the voice radio technology of that time. Issues that had to be overcome included developing

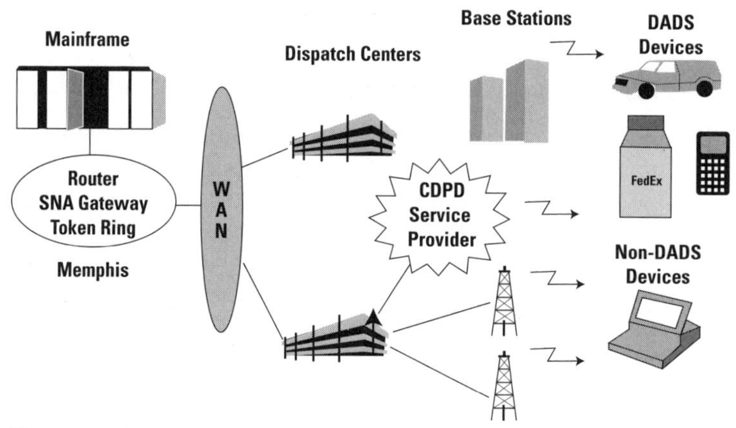

Figure 4.2 FedEx Network Overview

ROI: The Key to Productivity

FedEx has cited the following key benefits of the DADS system yielding Return on Investment (ROI) to the company:

- A reduced number of dispatchers (compared with voice dispatching) and more efficient use of on-road couriers.
- On-call pickups for customers.
- Uploads of package scans for near real-time package tracking.
- Package delivery service levels that are the highest in the industry.
- The ability to grow the courier force very rapidly in the early years of DADS.

protocols for the airlink, interfacing data modems over narrow-band voice channels and tuning RF systems for efficient data transmission. In 1980, the company's systems' 4.8 Kbps channel rate was considered very aggressive and was considered state-of-the-art for many years. "Our current 19.2 Kbps is still on par performance-

wise compared with public carriers offering similar capabilities," Winn Stephenson said. "One other challenge involved working with the FCC during the early implementation of DADS to ensure regulatory compliance of our system, since there were no rules and regulations governing mobile data over public airwaves at that time."

The Future: Public Networking, Hybrid Wireless Voice & Data

Although as of August 2001 FedEx maintained a private proprietary data network, "public wireless networks are clearly our direction in the future," Stephenson says. Up until then, the public networks have offered "not quite there" availability and reliability. "But we want guaranteed response time, and the costs have to come down. Increasingly, though, public networks have already become competitive with private networks, so over time we'll be moving to a public network because it is more economical." (FedEx Ground, a subsidiary company, has already made the move to public networking.)

Stephenson points out that the FedEx network actually carries a minimal amount of voice. In the future, FedEx will likely avail itself of the hybrid voice/data capabilities of 2.5 and 3G public wireless networks; so that drivers today who carry their own private cell phones or still call the company through landline payphones will no longer need a back-up system. Public networks could potentially reduce the costs of networking and ease integration with both wireless and wireline logistics, transport, and customer-related functions. FedEx will continue to improve its business operations and use new technologies to satisfy a tougher and more competitive package tracking and delivery system. As the first to innovate with wireless, FedEx stands poised to bring the next phase of robust 2.5 and 3G networking to the private enterprise space.

Northeast Utilities: GPS Mapping and Dispatch

Business Challenge: Rebuilding an RF System From Scratch

The first time Northeast Utilities (NYSE: NU) attempted a $1 million wireless data project to handle GPS mapping, untethered dispatch, and vehicle tracking, project outcomes did not reflect the technical team's expectations.

"We began a project in 1995 to pilot three new technologies," said Andy Kasznay, a software engineer who managed the wireless project for the Berlin, Connecticut-based energy company. "We wanted GPS real-time vehicle tracking – on-line projections of real-time information as to where our repair trucks were located. We also wanted wireless dispatching and for all these systems to be tied together." Northeast Utilities, which has almost 7,000 employees and services nearly 1.8 million electric customers in CT, MA, and NH and 187,000 natural gas customers in CT, needed an electronic, wireless version of a dispatch facility. Known as a "mimic board," this wall-mounted board illustrates how the different electrical substations are wired together. The board uses tiles or colored tapes to pinpoint locations of utility vehicles. "The first system took a year and a half to implement, and its performance fell short of our expectations," Kasznay said. IBM had put the initial wireless offering together, which included a combination of Bell Mobility computers, a Motorola radio with Data Radio modems, and mobility software built by Aveltech, a Canadian company. The system was based on a single-tower architecture, which proved to be unscalable.

"As the project went forward, we ran into a number of difficulties," Kasznay recalls. "The computer hardware chosen proved to interfere with voice radio communications. We replaced them with new computers, eliminating that challenge, only to discover that the radio communications infrastructure, a single, central tower site to which all the mobile units communicated, did not support expansion. The coverage area footprint was very limiting, and from a software perspective, the applications built right into the mobile units were not designed for flexibility. As the project proceeded, we determined it wasn't a suitable solution for the company as a whole."

Solution: Multi-Tower Radio Trunking System with Roaming

Northeast Utilities moved from its early-stage plan to a much more flexible radio trunking system, EDACS, a proprietary mobile system provided by Ericsson. "This system allowed a mobile unit to roam effectively from one tower site to the next, providing scalability and the capability of using the same bandwidth for different applications. Northeast solicited Requests for Proposal (RFPs) and selected Broadbeam to handle data transport and middleware development for the new wireless private radio network. Sybase was chosen to handle data storage on the mobile devices, helping link the application to a database. Kasznay and his staff wrote the custom applications needed on top of the Sybase/Broadbeam infrastructure.

The applications included the following:

- Facilities Mapping showing circuit locations throughout the company's service area.
- Street-Lighting application for dispatching streetlight repair trouble tickets.
- Environmental Spill-Reporting, used for emergency situation assessment, cleanup progress monitoring, chemical lab testing, and communications to appropriate agencies.

How the Application Changed the Business

"The spill reporting application now is no longer tied to a particular person," he continues. "It holds the information from every source, and is available to the entire staff, including chemistry people and the clean-up crew. Everyone involved in Spill Response is now looking at the same data, and it's enabled via the EDACS wireless infrastructure. Our system has gone from a manual approach to a technology-enabled one. People can go in and continuously update and review the data. It's definitely paying for itself over the paper-based system in both clean-up efficiencies and cost savings." Kasznay says each ruggedized laptop costs approximately $6,000, and each month of airtime per user costs approximately $30.00.

"Our mapping application is also important," Kasznay added. "In the past, we used paper maps, which were bulky and quite heavy. In order to transport those maps to work sites, we had to buy customized vans with specialized, heavy-duty springs to handle the extra weight of the maps. Carrying such weight, trucks could only cover a certain geographic area. By putting the maps on computer, each truck can respond to calls over a wider area. Because it's an electronic search, the drivers are able to pull up the information they need, respond more quickly, and complete their work in less time."

Return on Investment

Northeast Utilities calculated the savings associated with going to an electronic mapping application through the consumables saved. Over a five-year period, 560,000 D-Size maps would not have to be printed. This translates into 38 tons of paper, and Northeast Utilities estimated a savings of 1,800 toner cartridges. Over five years this was an avoided cost of over $400,000. The environmental savings alone were enough to justify the electronic system.

Lessons Learned: Future Proofing the Applications & Networks

Kasznay believes the most important lesson learned through the early implementation efforts was to "start small and scale up slowly." For example, the spill-reporting application enabled Kasznay's IT staff to consider sending power-outage reports to executives via smart phones as well as making wireless dispatch assignments to roving crews repairing streetlights.

Through these initiatives, Northeast Utilities learned about the importance of wireless scalability. "The early system was limited, and upgrades required costly and time-consuming software development," he said. "From a software and radio perspective, there have been obvious limitations in network coverage. In spite of the advantages inherent in the new EDACS system, bandwidth limitations and use presented some serious challenges. As public data network costs have declined dramatically, NU has turned to CDPD to supplement the network coverage."

With costs serving as a significant driver, Northeast Utilities has largely migrated to CDPD. "We still use EDACS infrastructure, but in the 1996 time frame we had built 13 tower sites. In our system we have so many radios used for many different purposes that capacity is limited. The cost of increasing the number of channels per tower site is less favorable than CDPD adoption. It is now cheaper to buy a modem from a CDPD carrier and pay monthly usage charges than it is to build another tower site, so we use a combination of both public and private networks." Kasznay says that CDPD has a number of significant benefits: cost, speed (19.2 Kbps currently), and latency. But because no network is perfect, the infrastructure is designed to operate transparently to a field inspector during radio coverage 'dead spots.' The information updated during the "downtime" is cached until a wireless connection is reestablished.

"Right now we have 120 people using the wireless system, and we have applications in replication mode, including docking, which enables users to update information to their computers in the evening. The heaviest concentration of users now is

Bottom-Line Benefits

The key benefits of Northeast Utilities infrastructure are:

- The ability for only changes in data to traverse the wireless network, reducing airtime costs.
- The ability to have all the data the user needs on the local machine so that it can operate in dead spots; hence, the system's response time is not dependent on the network.
- Infrastructure support for multiple networks, enabling future upgrades.
- A familiar applications development environment, enabling developers to write (i.e. Visual Basic/Sybase) without the need for addressing the complexities of wireless networks.

among utility employees in Connecticut and Western Massachusetts. Kasznay's advice: *Involve users along with software developers in the wireless planning process and give bi-directional feedback.* "That was a key to making it work the second time around," he said. "For the last four years, we've been adding successful applications, and we expect to continue to do that with handheld wireless devices. There is plenty of work to do."

Telemetry and Wireless Troubleshooting

Business Challenge: Messaging, Control, and Data Collection Between Machines and Humans

The two previous case studies involved human-to-human communication and have showed that improving the communication flow between mobile users and their non-mobile counterparts can deliver strong business benefits when enabled with wireless data applications. Another value chain that can be improved through the application of wireless data is the sharing of information from machines in the field, such as vending machines (inventory control and product freshness monitoring), vehicles (truck driver performance or passenger car repair information that can be sensed by a central repair site), or appliances (remote diagnostic tests). Technology advancements in devices such as appliances and cars have placed microprocessors, memory, and other computer pieces directly into these machines. Machine designers are becoming aware that they can use the information gathered by these embedded computers to improve machine performance or to avoid machine repairs. Typically, this is done by storing performance data of the machine and using remote diagnostic tools and/or human experts to analyze the data to predict and avoid machine failures or to send tuning parameters to the machines. The range of applications is growing and is a natural area for future growth given the richness of digital data now captured and available in machines. A truck can easily report its vital statistics (engine speed, miles per gallon, number of miles driven since last oil change . . .) to a company's home office so that fleet maintenance can be simplified. Likewise, a vending machine owner can sense when each vending machine needs inventory replenishment or if the vending machine is in need of repair.

Enterprises are finding many ways to use remote sensing and automatic meter reading (AMR) – part of an explosion of applications under the general heading of wireless telemetry. Yankee Group projects that telemetry equipment service revenues will be a $6 billion market by 2004. Today, over 6 billion embedded microprocessors are in operation worldwide, providing controls and monitoring functions for everything from refrigerators and vending machines to oil wells, delivery trucks, and alarm systems. Moreover, for every piece of monitoring equipment, there is a company that needs to monitor, maintain, or adjust settings on the equipment.

Telemetry grows out of technologies that utilize over-the-air transmissions to handle various forms of aging, messaging, and control between machines and, quite often, machines and humans. A number of proprietary data networks – both satellite and terrestrial – have been devised for wireless telemetry, and applications written for these networks are now using remote sensing, alarms, and notifications via the World Wide Web. Some telemetry systems rely on AMR based on next-generation Low-Earth Orbit (LEO) satellites that integrate Global Positioning Systems. The satellites determine the precise location of remote devices. Further, AMR has become a tool and "cash register" for the utilities industry. It is "disruptive" because it mines data from machines and meters over-the-air, altering the time frames, the level of control, and the effectiveness and speed of interaction between systems and customers. For example, Web-based telemetry is now being used to send automatic shut-down notices (which enable a controlled loss of power) to hundreds of grocery store refrigeration systems in California, a state beset by energy shortages and rolling blackouts.

Wireless telemetry can provide a competitive advantage by delivering customer usage data on a daily or hourly basis. In the case of energy management, the data dumps directly to a utility's billing and customer information systems, enabling the company to identify usage trends, reduce billing costs, and target marketing campaigns to specific kinds of customers. Most important, telemetry reduces the need to mobilize human technicians to do perfunctory meter reading. It provides polling, paging, and automatic alerts to field force technicians and staff when needed.

Solution: Pervasive Machine Telemetry, Remote Messaging, and Monitoring

One of the industrial strength RF networks designed for AMR is Aeris Network's Microburst, which utilizes a proprietary method for transmitting short messages (i.e., data packets) across the control channels of cellular networks (See Figure 4.3). Using standard IS-41 protocols, cellular devices accessing the Aeris network can transmit short data packets on underutilized channels, allowing the growth of a variety of new messaging applications. When combined with Web-based servers, Aeris can enable remote two-way paging and alarm control for air conditioning and refrigeration systems. A new level of energy management and cost savings is the result.

Notifact ™ of Fairfield, New Jersey, a leading player in automated telemetry, provides 24-hour message routing, remote monitoring, and service solutions for heating, ventilating, and air conditioning (HVAC) equipment. The company has also developed systems based on wireless Web technology. The company utilizes the Aeris Microburst network to provide near ubiquitous national coverage for its Web-

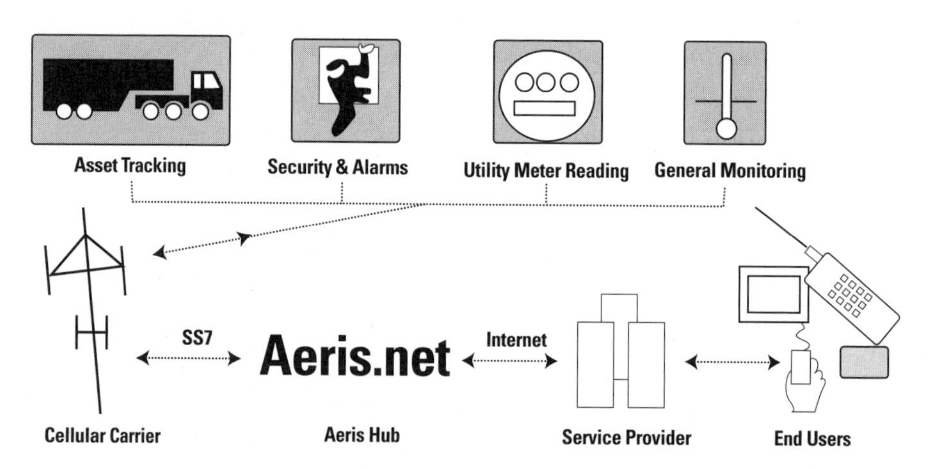

Figure 4.3 Aeris.net's Network: How Microburst Works

enabled two-way messaging system between machines and humans that manage the machines. Further, it is expanding applications to include telemetry for standby power generation, monitoring of fire pumps in buildings and dumpsters (systems that signal when dumpsters are full, thus reducing the number of trash collections and costs of hauling), and remote control of refrigeration systems. Notifact, in short, is in the business of creating intelligent human/machine messaging.

Current Configuration: A System for 'Universal' S.O.S.

The Notifact system works in conjunction with the Microburst network and the Internet to provide telemetry service. A simple example: when an air conditioner or heating unit needs maintenance or repair, it fires off a wireless S.O.S. via the Notifact transceiver. Using MicroBurst's North American AMPS network, the transceiver forwards the message to the Notifact Network Operating Center (NOC), which then relays it to the appropriate person by e-mail, fax, phone, pager, or even by XML message into a service dispatch system. The system works on the principal of "live messaging," which means it will not leave voice mail messages to key personnel, but will require them to respond to alarms by inputting a digital pass code (only then is the message considered to be delivered).

Originally the system was developed to provide a wireless replacement for wireline-based machine paging and monitoring. "Wireless (signaling) to machines addressed a lot of problems companies faced with the installation of phone lines – such as the problem of unauthorized access (to machine information)," said David Sandelman, Vice President and Chief Technical Officer at Notifact Corp. "Wireline telemetry guided us toward utilization of wireless transport; but with that, we decided that the utilization of servers and the Web site would give our customers a lot more power and ability to send messages to various places," he said. In other words, "no longer would a machine just be calling a modem with a line printer. Instead, Web-based telemetry would afford us the ability to send messages to pagers, faxes, e-mail, and telephones," thereby extending the reach and efficiency of the messaging.

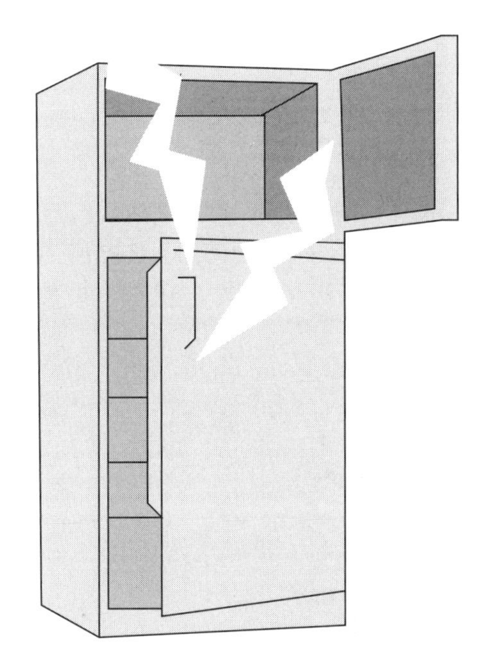

Why is this important? "Let's say a restaurant has an air conditioner that suffers a compressor failure. No one may become aware of the problem until the temperature begins to get uncomfortable and diners start arriving. An hour or two may have passed to identify that there is a problem and perhaps another few hours to get someone to service it. In the meantime, diners are leaving for cooler climes," Sandelman continues. "Notifact will alert the restaurant owner to the problem by e-mail, or send a fax to the manager, or both within minutes of the failure. Simultaneously the alert message, which contains the precise nature of the failure, is sent directly to the person most qualified to fix it. In the past if your A/C went down, you would call the contractor. But with RF telemetry, the service technician could already be working on the problem before you even know about it and reach for the telephone to call."

Web-Driven Messaging

Notifact uses MicroBurst technology to transmit information over the public SS7 cellular network. The technology was selected because it offers a major advantage: ubiquity of coverage (See Aeris.net coverage map in Figure 4.4). "Prior to Aeris's debut in early 1998, we were looking at other transports like CDPD," said David Sandelman. "But even prior to CDPD's commercialization, the issue was coverage," he said. "We needed to offer a wirelessly enabled product across the board without restrictions as to where it could be deployed, and systems like CDPD were covered only in particular markets, and this led us to Aeris." Nonetheless, Sandelman notes that "our application was written in such a way that we're not tied to any one transport." This is significant for enterprise customers; an effective wireless application may require multiple networks and transports to provide customers the coverage they need.

Aeris' high-bandwidth solution enabled Notifact to send more information to customers – about 1 Kilobyte of data. "When a piece of equipment sends a message, the information is carried by the Aeris system to the Notifact message center where it is forwarded based on the user's configuration," Sandelman explained. "The user has already configured at the website where and how messages are to be delivered. Options include whatever medium is most appropriate for each recipient or situation. If the equipment sends a high temperature alert, the system can send e-mail message alerting the appropriate person. If the equipment's temperature is too low, the message may reach its destination by fax. Both messages can be simultaneously sent to a third party." Since the software needed to operate the system resides at the website (www.notifact.com), the user can access and change the system using any ISP from anywhere in the world.

Further, the system is configured to send a "heartbeat message" to inform the user that the unit is still working. Along with the heartbeat is a recap of the

Figure 4.4 Aeris.net's Microburst Coverage Map

equipment's performance the previous day – how many times the equipment cycled, for example, and for how long. The user can look up this information at the website and read up to 90 days of history – an aid in detecting problems with performance. The data might show, for example, that an air conditioner that is supposed to turn itself off on weekends remains active. Knowing this can save a property owner gratuitous energy expenses.

Another Application: Mass Broadcast Telemetry and Control of Lighting Systems

More sophisticated applications of two-way telemetry are on the rise. For example, Notifact's two-way systems are now providing "mass broadcast" telemetry messages to lighting systems in supermarket chains. In California, for example, the Public Utility Commission (PUC) is offering rebates to certain customers to turn off their lighting for a limited time (e.g., a couple of hours) to save power and avoid rolling blackouts. With telemetry, supermarket chains can now get a rebate for installing automated telemetry systems, cashing in on their "downtime" from the electrical companies. "We're able to send a command instantaneously to turn off power use in a hundred grocery stores," Sandelman said. "The units will be installed in hundreds of California grocery stores." Making a single wireless call to 200 sites or more (rather than 200 separate wireline phone calls) yields enormous efficiencies. "If every store had to log onto the Web using a conventional dial-up modem, they couldn't do it within the thirty minute window mandated by the CPUC," Sandelman explained. Under the new model, a store manager will get an automatic notification that the lights will be turned off, and with that store chain management will implement a shut down of their stores' lighting across the affected areas of the state. The shut down is implemented via the Notifact Web site; this initiates a sequence of events that includes an e-mail or page to store managers, warning them that their refrigeration units will be shut down for a two-hour period. As the sequence progresses (e.g., within the next 20 minutes), the system generates an automatic signal to Notifact transceivers, causing the system shut down.

Return On Investment & Lessons Learned

For many years industries have kept track of the health of their capital equipment with the help of automation and energy management systems that utilize dial-up telephones and a central monitoring facility. These highly complicated technologies work well in an industrial context, but are inappropriate for a non-industrial environment employing smaller pieces of equipment because of high capital costs. This is where wireless telemetry is most effective and economical. Web-based telemetry systems are much more compact and utilize both server-based applications software and RF signaling to reduce capital cost.

For instance, rather than wiring an entire building automation system for alarm signaling, contractors can buy separate telemetry devices (including the Notifact transceiver), which they can install themselves for $1,000 per unit compared to a minimum of $10,000 for a wired building automation system. Alternately, contractors can purchase telemetry systems pre-installed in major OEM manufacturers' air conditioning, heating, and refrigeration systems (e.g., Lennox, Solidyne, Fireye, Com-trol, a division of Invensys). Since the cost of the Notifact system is less than ten percent of building automation systems requiring specialized PCs, programming, and software, the return on investment is substantially greater.

Of course, measuring the savings by avoiding business interruptions is very difficult. "The challenge now is who pays for the system when all parties benefit: contractors increase service business and efficiencies, owners improve peace of mind and tenant service, and equipment manufacturers use the telemetry device as a way to gain competitive advantage and grow equipment sales," Sandelman explains. "The biggest lessons we're learning right now is about inertia – how long it takes contractors and OEMs to change the way they've been conducting business."

The Future: Pervasive RF Telemetry

Wireless telemetry is well suited to an untethered society. It can monitor pumps that run municipal sewage and water supply systems where running land lines could be both expensive or impractical. Other applications include the monitoring of hospital equipment, cold storage facilities, green houses, and zoological environments containing heat- and cold-sensitive animals such as reptiles and amphibians, equipment shelters, and even the beacon light on top of a radio antenna tower.

Consumers may also benefit from telemetry and residential gateways (see Chapter 5), especially people with second homes that remain unoccupied for most of the year. Unless they have money to burn, absentee property owners worry a great deal about their investments, and the grim possibility of boiler and heating system failures and frozen pipes that can burst.

"When homes become automated we will enter the age of the smart appliance," Sandelman explains. "The machines will be equipped with wireless monitoring devices, and people will get paged at work when their refrigerator detects a problem that needs fixing. At first the systems will work mostly with major appliances such as washing machines and refrigerators, and the heating and air conditioning systems. But it won't be too long before the technology drops in price and by then we could be talking about telemetry for coffee machines and virtually all electronic devices," including our on-board control systems in cars. Telemetry and wireless communications will become inextricably woven into the fabric of our lives.

Designing an Effective Police Data Radio System

Public Safety Challenge: Better Cops, Criminal Detection

The City of London, Ontario, is large by Canadian standards: 340,000 people. The Police Force has been designing a wireless data solution to improve criminal detection, records management, and mobile information transfer since 1993. Prior to that, the police force of 468 officers and 168 civilians were using voice radio dispatch. By March 1993, the Police Services Board began a four-phase project to develop centralized computer records accessible via wireless, as well as computer-aided dispatch, and wirelessly enabled data capture and image transfer capabilities to and from police cars.

Before the conversion into a wireless data radio system, the police officer had to hand-write all incident reports, turning them in at the end of their shifts, which could run as long as ten hours after an officer started work. The handwritten materials were audited and passed to data entry to upload into a central computer, resulting in major delays. The police department was up to three weeks behind in data entry. In the field, all requests for information were called in by radio; if the information was available on the central computer, it was relayed back – verbally – to the officers. But information could be slow in coming back to the field, depending on the backlog of requests for the operators. Moreover, the old systems had no capability to scan, send, or receive images, making it difficult to verify identities of suspects during arrests.

Solution: A Secure Mobile Trunked Data Radio System

"We needed to replace our 20-year-old radio system that would do only voice with a complete data radio system communicating with mobile workstations in all patrol vehicles," (see Figure 4.5) explained Eldon Amoroso, director of information and technology at the London Police Service. After considerable research, the Police Department contracted with a Canadian systems integrator, Versaterm, to provide the software managing the main records system controlling dispatch, scheduling, and criminal records. In addition, the Police Department adopted an Ericsson EDACS 800 MHz mobile trunked radio system to provide secure wireless network capability.

Figure 4.5 London Ontario Police Use Wireless to Improve Efficiency

Installed in 1998, the private trunked system supports 1.5 million radio packets a month in addition to voice. The Police Department transmits data at 9600 baud over 17 channels, using an automatic switching system. "Versaterm was very helpful in telling us what questions to ask in our RFP," Amoroso said. "For example, we specified how many transactions at 100 bytes (roughly the size of incoming transactions used when querying a name or license plate), and responses from those in thousand byte units, and how many 10,000 bytes required for mug shots," he said.

Early Stage Lessons: Building Ironclad RFPs

Up until Amoroso began ferreting out requirements for data, most vendors were relatively casual about it. "In the past, radio companies tended not to want to specify hard deliverables on the data side," he said. "For example, when Ericsson came in to the project, the general attitude was 'Don't worry about data, we do it.' But our criteria from the IT side was more stringent; we put in performance data throughput requirements that the vendor would have to meet. We also talked to other police departments that hadn't put in a spec for their systems – and they regretted it. The systems didn't work as required."

Indeed, modeling the throughput and demanding specified levels of EDACS capacity and performance was one of the best moves London Police made. "Nobody we knew was doing that volume of data in Canada at the time," he said. "Certainly people weren't doing mug shots or uploading reports on the street. Initially, when we sent out the RFP, Ericsson and Motorola came back to us and said our data estimates were completely unrealistic, but we said, 'You've got to look at what we'll be needing five years from now.' Even though we knew we couldn't do mug shots right away, we wanted the bandwidth. And the officers in the street wanted more and more in their cars. We did a little survey of the officers and discovered we had underestimated how much they loved the wireless systems. They were absolutely sold on them; especially with the availability of downloading mugshots for positive identification, it was making their jobs safer."

A Critical Choice: Wireless Middleware for 'Network-Agnostic' Operation

The RFP that resulted in the EDACS contract did not invoke a CDPD network service option because CDPD at the time was unavailable in the London region. However, police IT wanted to build a private radio system that could eventually adapt to changes in wireless applications, network options, and protocols. "We had all these applications we wanted to run over EDACS," Amoroso said. "So there were two alternatives: either write the applications so they talk directly to the radio system; or much better, use middleware that sits between the application and the wireless network, actually shielding the app developer from the vagaries of RF networks and their radio protocols."

"Middleware isolates the application from the radio network," he explains. Using Broadbeam Corporation's middleware library written for EDACS, the London

Police Service and Versaterm were now free to develop mobile applications without having to worry about the wireless transport. Versaterm had chosen to partner with Broadbeam to provide the wireless development tools for the EDACS network.

"Let's say we need a lot more data capacity in the future; we could buy the CDPD middleware library, and we could be transmitting CDPD data (under the same applications) that afternoon with no application changes," Amoroso said. "From the business point of view, middleware enables the customer to manage the radio communications and buy whatever services and products they need without worrying about future radio system additions or modifications."

Current Configuration

At present, the EDACS system is serving London Police Department needs extremely well. "At any given time it's not unusual to have 80 cars on the street, and we have portable units – both laptop and radio," Amoroso reports. The Police Department first used NEC 2200C laptops connecting to Windows NT Servers and Compaq UNIX Server for the central computer system. However, the mobile laptops in police cars are changed every three years; new models include the Panasonic CF 27 ruggedized units. "We've raided chop shops, actually gone there and used the mobile workstations to do queries to the national system on stolen vehicle license numbers and serial numbers. We've been able to get color mug shots and squeeze them down to 4K using compression techniques." The entire system is extremely effective, he says.

Figure 4.6 Wireless Patrol Car

Broadbeam's solution has also allowed the Versaterm-authored mobile application software to run without interference over the EDACS network. In addition to writing the middleware, Broadbeam's Professional Services Group was contracted to develop the interface to Ericsson's CADLINK II status message gateway. The middleware compresses data and optimizes acknowledgements needed to send and receive information, thereby improving network capacity.

Figure 4.7 London Ontario Police Wireless Task Force and Mobile Workstation Task force includes (from front to back): Director Eldon Amoroso, Superintendent Rick Gillespie, Andy Hunter, Staff Seargeant Peter Glen, Brad Resvick, Annette Swalwell, Acting Chief Brian Collines, Constable Bob Plows, Andy Bennett, Bill Berg, Wendy Gaffney, Superintendent Brad Duncan, Bob Aves, Jan Smelser, Ryan Holland and not shown: Jeff Craigmile, Randy Van Puyenbroeck, and Case Huysmans

The London Police Force now has 85 fully equipped cruisers (see Figure 4.6) with the new mobile communications system. Each car has a mobile laptop (see Figure 4.7) that communicates with an NT server in the station; this connects to a Compaq UNIX server, which houses the central computer system. Information is transmitted using the Ericsson Orion Radio over the EDACS 800 MHz trunking system. Approximately 250 of the 430 London police officers use the mobile solution in the field. The costs in Canadian dollars were $4.5 million for installing the radio system; approximately $1 million for the mobile workstations; $1 million for the application software, and $100,000 for the middleware. The force pays Ericsson $120,000 annually for EDACS network maintenance and services.

Lessons on the Street: Safety and Faster Response Times

The complete system allows officers literally to take their office on the road, enabling them more active working time. The laptops are used to take witness statements, write incident reports and narratives, and make queries back to the central computer system and the National Police Information Networks. The laptops can be carried when the officers leave their patrol vehicles to respond to a call; the laptops then relay all of the information gathered at the scene when the officers return to their cars. The system also has a safety element. Officers can make a query on a license plate or suspect before approaching the vehicle; and if a police cruiser is out of contact after a certain period of time, automatic "timers" are set within the CAD system to send alerts to dispatchers. Further, the mobile units are configured with an interface to the EDACS control channel, allowing emergency signals to be instantly displayed at the Control Center with officer identification and last location and activity information displayed. Each officer is equipped with a portable radio, which has an "emergency pin" that can be activated to enlist assistance.

The Future: Untethered Police Everywhere

The London Police's wireless implementation is a pioneering example that has set the precedent for many other Canadian and American police forces. "We won the National Award for Technology Implementation," said Eldon Amoroso. "The officers on the street want more and more technology in the car (for example, dual officer teams requested two individual computers in each car). The officers are absolutely sold on it; it's been a wonderful tool for them, and is making their jobs safer."

Bottom Line Benefits

The operational benefits of the wireless network include the following:

- Less paperwork. The force has been able to relocate in excess of 30 percent of its central record staff, and the London Police believe the system has increased the force's exposure in the community without adding officers.

- Better records management and faster information reporting and turnaround. The police department now captures 100% of the text from the field with information becoming available within minutes from the time the officer uploads it. Management can easily monitor investigations and ensure timely disclosure to the prosecuting office. In some instances, the system expedites investigations and subsequent arrests.

- Voice traffic has been reduced on the network. London police report they've experienced a 30 percent reduction in voice calls in the first three months that the data facilities were in operation, dropping from 350,000 voice calls to 230,000 calls per month.

The technical benefits of the middleware-driven system include the following:

- Optimized transport eliminates expensive network overhead and substantially reduces packet sizes, thus minimizing the radio time required to transmit data.

- Higher network capacity. More information can be transmitted on the existing network than would have been possible without using the middleware solution.

Bottom Line Benefits — Continued

- Middleware "shields" the application from the vagaries and differences between wireless networks and protocols. Enables virtually instantaneous start-ups on new radio systems.
- Middleware automatically adapts transmission to "fringe" signal conditions, ensuring optimal performance. If data is lost, the system will resend only the packets needed to complete the transmission once radio conditions improve. Moreover, the mobile and server applications do not have to 'understand' or adjust for these conditions because middleware handles the nuts-and-bolts interface with the radio system.
- Data traffic is handled economically. By December 1997, more than 700,000 packets of data were being transmitted each month, and the load was increasing. Broadbeam's middleware compression yields 3:1 character reductions in text messages. Without this compression, data volumes monthly over the data radio would exceed a gigabyte.

Fast Auto Claims Estimating

Business Challenge: Enabling Insurance Adjusters to File Auto Claims On the Spot

Auto insurance estimators now have a reliable wireless system to do auto claims processing at the scene of the accident. ADP Claims Solutions Group's Pen-Pro® mobile estimating system is now integrated with ADP's ClaimsFlo® Wireless, a product enabling mobile insurance adjusters to "walk the vehicle" with a mobile terminal, filing an estimate and even cutting claims checks on the spot. This answers a huge efficiency challenge in the insurance industry. Wireless estimating is a "disruptive" technology because it collapses time frames for customer turn-around, makes filing more efficient, and reduces paperwork and "windshield time" an average of one hour per day per appraiser.

Solution: Eliminate Redundancies in the Internal Claims Process

Collectively, the ClaimsFlo system streamlines the appraisal process. Claims adjusters can now use wireless estimating and communication to access relevant data on customers and transmit claims information from the field or a drive-through service center. Appraisers using the PenPro mobile estimating system can download new assignments and exchange information with trading partners and staff adjusters. Claims information, including auto make, model, year, engine, and parts options, are transmitted wirelessly using Itronix 6250 computers equipped with Windows 95 and Windows NT.

Broadbeam Corporation provides the middleware, enabling the ClaimsFlo Wireless application to operate on PenPro mobile wireless systems. Information from the insurance appraisers' ruggedized computers is transmitted on the Mobitex network provided by Cingular. Today, COUNTRY Insurance and Financial Services (Bloomington, Ill.) is using the mobile data estimating system among more than 100 material damage appraisers in nine states.

Early Stage Objectives

Before the year 2000, COUNTRY Insurance and Financial Services had been using an award-winning PenPro mobile estimating system to enable appraisers to do their work "on-site," pulling up extensive information. The majority of the estimates generated by appraisers out in the field would then be sent back to the home office via a modem for processing.

That changed when ADP Claims Solutions Group (CSG) incorporated a new element to PenPro's estimate process with ADP ClaimsFlo Wireless, which allows appraisers to transmit claims information from the field in real-time, significantly increasing the levels of customer service and shortening response times and turnarounds for claim checks and repairs. Bypassing traditional modem communication, appraisers save approximately one hour in paperwork and drive time each work day to complete and receive assignments, estimates, and total loss evaluations wirelessly. ClaimsFlo Wireless integrates all of the ADP mobile estimating systems together, enabling insurance adjusters to disseminate and receive data instantly from trade partners and staff adjusters. The insurance appraisers now use wireless IP to settle claims on the spot.

Configuration and Network Set-Up

Currently, the system uses the Mobitex network provided by Cingular Interactive with middleware provided by Broadbeam. "Smart IP® enables the wireless connection to look like a standard IP connection to the user terminal," according to Anthony Esposito, a vice president of worldwide sales at Broadbeam Corporation. This in turn is complementary to the IP-based claims processing system. Field adjusters are equipped with Itronix computers and built-in RF antennas. Walking the vehicle, they create estimates using PenPro, and send the completed estimate data to ADP's host computer over the Cingular Interactive network. Smart IP mediates the wireless connection to the legacy system, enabling 'network agnostic' communication over Internet and private networks, thus reducing airtime expenses. Before the appraiser gets down to an accident, he can download an assignment. Onsite, he can prepare the estimate and file it wirelessly." ADP sells this application to insurance companies. COUNTRY Insurance and Financial Services is among the first to implement the wireless portion of the system for appraisers.

ROI, Lessons Learned, and Future Functionality

COUNTRY Insurance and Financial Services reports that appraisers are saving about an hour a day in claims processing since they no longer have to go into the office to upload or download claims data or to receive assignments. ClaimsFlo Wireless guarantees the company that insurance information is being transmitted quickly and securely and that appraisers are making decisions with current information from the company's data stores.

Recently in Chicago, a customer's car was vandalized while parked at a White Sox game. The customer had reported the claim the following morning on the phone and was told an appraiser would try to see the vehicle later that day. However, the appraiser Ken Smith was able to get to the site that morning by picking up his assignment wirelessly in his car. Smith went directly to the location, met the customer, inspected the vehicle, completed his estimate and issued a draft by 10:30 AM. Smith says that the wireless connection has paid off many times. He's had customers ask, 'How did you get here so fast?'"

As network coverage expands, COUNTRY Insurance and Financial Services plans to roll out the wireless service to all of its 150 appraisers. And using middleware, ADP CSG will add support for complementary wireless networks, such as satellite, as well as new devices in the near future.

5 THE WIRELESS CONSUMER: WHAT WILL IT TAKE TO WOO THE MASS MARKET?

Spreading Network Intelligence like Wildfire

IN THE HOUSEHOLD OF THE FUTURE SET IN Hertfordshire, England, it is snowing as perfectly as a dream. The snow is the kind you see when you shake up a little glass bauble and raise a "storm" of winter white. The flakes cloud a view of a modern "wire-free" British house and children with frozen smiles.

The children, boy and girl, are looking out of a big picture window at flakes as fine and white as needlepoint. "Wildfire," the Father calls out, "lights on ... tellie on." It is early morning as Father and Mother snuggle in their warm bed. Mother rises slowly, kisses Father, stretches, asks Wildfire to check the baby's room; a tiny white ceiling-mounted video camera provides an instant picture on the TV screen of a warm, comfortable baby rubbing eyes full of sleep. "Wildfire, turn on the coffee percolator." "Done," says Wildfire, in the voice of a cheerful English nanny. Wildfire now turns on the stove; the washing machine; regulates the thermostat; entertains the young boy at his Internet screen in the conservatory; retrieves e-mails, Web clips, and messages; adjusts lighting; checks the outside doors; and monitors the father's heart during his workout routine. In short, this wireless servant ushers the house via telemetry, image, and speech recognition toward a full, organized morning wakening.

"Wildfire."

Wildfire is an invisible agent, a trusted presence that spreads throughout the fabric of these British lives. Not exactly a robot, but something like it, the "presence" is a wire-free intelligence within this house, which has ears to listen and eyes to see, but no arms and legs. Wildfire actually is a system of sophisticated interfaces, networks, messaging protocols, and speech recognition subsystems. It recognizes the voices of each family member, controls and monitors all communications, utilities, appliances, access to the Internet, CATV and security functions, and more. The Wildfire network that "lives" wirelessly within the house is accessible via fixed and mobile devices equipped with microphones. Each component within the intelligent house (see Figure 5.1) communicates with the rest of the system using a Java-driven messaging protocol developed by Sun Microsystems. Wildfire is virtually omnipresent, in constant touch with the world of the family and the outside world. The network boasts high bandwidth, mobility (even while humans stand still), over-the-air activation, text to speech conversion, and command-driven access to both local and wide area networks, even while family members "roam" in their cars – and the cooperative style of a well-mannered English woman.

"I'm here," says the voice.

Fig. 5.1 Wildfire House of the Future

Father now checks the groceries and settles at his desk, reviewing his daily schedule and making travel plans over the Internet. "Wildfire." "I'm here," says the voice. "Order the basic shopping list, he says, sipping his tea. "Wildfire, what is my schedule?" The Wireless assistant accesses Father's calendar, contact lists, buyer's inventory, and enterprise databases. For Mother, Wildfire runs the residential gateway to household appliances and fuel pumps, carphones, and GPS displays, providing access to banks, brokerage houses and Web sites, her own job and corporate intranet, as well as communications, and monitoring of the children's rooms. Wildfire "follows" Mother from home to car to work and back again; part of the "extended reach" and intelligence of the network.

Driving now, Mother talks to Wildfire inside her car, speeding along the slick dark roads as it starts to snow again. "Wildfire, Central heating, 24 degrees Celsius." "Done," Wildfire says. As Mother nears home Wildfire greets her peaceably once again by sensing her car's approach and automatically opening her garage door. How much of what we are seeing is real?

Lights, camera, action: Actually, what we've been doing is watching *a movie* about Wildfire. We now see scurrying cameramen and video equipment piled up outside the house. Camera pulls back to a wide shot of the house from outside, as though looking down from a helicopter. Snow blowers are pouring jets of phony snow over the house as the camera crew moves in and out of range to film the Wildfire project.

The film caption reads: "The only thing artificial in this film is the snow."

The audience – about 1500 people attending the Cellular Telecommunications Industry Association (CTIA) Wireless 2001 in Las Vegas – laugh nervously. It's a release – the wireless is real; the snow is not. So much for contemporary plays on reality.

The Pervasive Wireless Model

The Orange Intelligent Home Research Centre is a £2 million "home of the future," a test bed for wireless products and services, as well as solar panels, recycled water, and robotic lawn mowers. Orange – the company – is the one of the largest European wireless carriers. Acquired by France Telecom in May 2000, Orange is led by a flamboyant CEO and wireless visionary Hans Snook, who has guided the pan-European carrier on an in-depth examination of the untethered lifestyle – what Snook calls "wirefree ™ working." Roughly translated, "Wirefree is about [giving] people the freedom to communicate – whenever, wherever, and however they choose," he says. "Wirefree working is to help define and explain the fundamental change in technology and work practices that has been underway for five years or so," he continues. "It refers to the use of Orange products and services for doing business outside the traditional environment, whether as a member of a 'virtual team' scattered around the globe, a sales person on the road . . . or a teleworker who works from home some or all of the time. It is a broad concept and takes in such options as hot-desking, teleworking, and mobile working."

With more than 33 million subscribers in the UK, France (and French possessions), Europe, and worldwide, Orange brought about Wildfire much like Ray Bradbury created an autonomous robotic house half a century ago in his post-nuclear morality tale, "There Will Come Soft Rains," part of *The Martian Chronicles.* In this story, a robotic house, filled with voices and machine presence, act much like trusted wireless friends, cheerfully maintaining routines by doing the essential cleaning, cooking, lighting, and other chores around a house curiously abandoned by humans (most likely, humans vanishing in the nuclear holocaust). The trouble comes when systems short and sparks fly – the robots go haywire trying to interpret the puzzling silence of masters who would normally "talk their intentions." Eventually the robotic wiring and the house catch fire, and nothing but post-nuclear rain can quell the flames of a machine experiment gone awry.

In the Wildfire update – at least as Snook presents it – things are happier. The trusted "friend" concept incorporates the versatility of "3G" and ubiquitous human-wireless interactions. It includes the ability to "hear" and interpret voice commands, remain in 'always on' mode, translate text into speech, transact numbers and business, deliver wireless multimedia (voice, data, streaming video), on telescreens or computer screens, utilize excess voice capacity for short-range data transmissions, remotely monitor children and entryways, and control energy and communications systems.

The Connected Home of the Future

Though it is never explicitly stated in the film, Wildfire is a completely integrated, pervasive system (Orange calls it an "integrated intelligent home"). It is as connected as the traditional office environment is today and then some. The home allows access to TV, movies, and the Internet from any room, enabling family members and even things (e.g., machine and utility monitoring systems) to communicate with work, home, friends, offices systems – whatever they require. The numbers of technologies involved in creating the intelligent home are astonishing (See Table 5.1 for a list of device and technology suppliers). More than 50 leading edge suppliers and partners such as Ananova, Dyson, Sony, Tunturi, Compaq, Intel, LM Solutions, Newstakes, Nuance4, and SunMicrosystems have supplied a range of advanced systems and appliances. Wirefree devices and systems included Orange phones using Wireless Application Protocol (WAP) browsers and SMS; wirefree PDAs/Webtablets with push button access to utilities, heating and appliances, wall panels in the house, and Web-connected PCs. The home's core software (see Figure 5.2) is run by Sun Microsystems' Java and Javascript (a high level, machine inde-

Table 5.1 Orange at Home Partners and Suppliers

Technology Partners	Sun Microsystems
	Compaq
	Intel
	Ananova
Consultants	SMC
	Urban Salon
	Woodman Construction Management
Technology Suppliers	Panja
Hardware/Software	Echelon
	Navaho Technologies
	Nuance
	Dyson
	Cobalt Networks
	ILOG
	Hitatchi America
	Unisys
Software Developers	IT Solutions
	WDS
Others	BearBox
	Fast Systems
	Royal Auping
	Seachanger
	Viessmann
	Movement Control Systems
	Black Box Telematics
	Tunturi
	Gaggenau
	Eco-Logic
	Solar Century
	Friendly Robotics
	Volvo

pendent programming language); JINI, (Sun's ad-hoc networking software, which uses a message-passing middleware layer between the software components of the core system); and Javaspace. The latter technology provides a distributed messaging capability and functions with JINI to enable different software components to communicate seamlessly within the home network despite being hosted on different hardware platforms (e.g., Windows NT4, Red Hat Linux, et cetera). All of the communications systems work in tandem with environmentally friendly recycling and energy/water conserving systems.

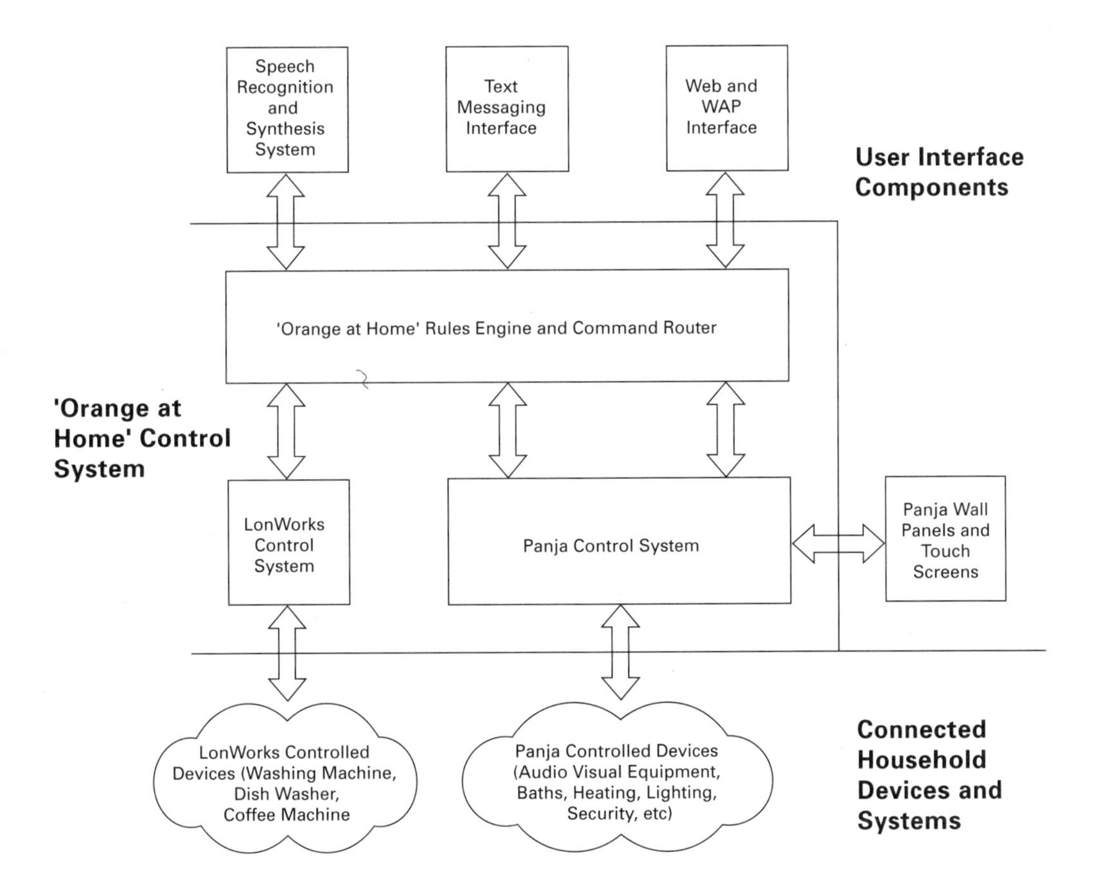

Figure 5.2 Orange at Home Control System

Wireless becomes a roaming accessory, an ethereal "ear and voice" of an untethered, society.

In totality, the Orange at Home pervasive wireless system is quite complex, even though it presents a "simple" and intuitive picture. Based on independent research, Orange is betting that by 2005, the market for "intelligent" homes will equal that of the PC market of today, and 1 out of 5 homes will incorporate intelligent wirefree technology, the company claims.

Judging from the reaction among CTIA Wireless 2001 mavens, Orange and Snook may be onto something. Wildfire exemplifies a far more powerful paradigm for wireless data usage than any presented to consumers thus far. The Wildfire concept points to *pervasive untetheredness* – much like pervasive computing – in which wireless becomes a roaming accessory, an ethereal "ear and voice" of a mobile society. Functioning independently of specific devices and terminals, networks, and gateways, Wildfire spreads. Its intelligence has ubiquitous presence in homes, cars, shopping malls, eateries, libraries, airports, bus stations, factories, schools, office buildings, and people's mobile desktops. Because it represents an "environmental" wireless model, Wildfire could mark the true direction for this century's wirefree consumer lifestyle.

Now, who will pay for it? Presumably, homebuyers will, by purchasing the communications and ecologically sound infrastructure as part of their "intelligent home" mortgage. *But is this really how wireless data technology will make itself available to the mass market?* Absolutely not, for while this is glorified version of some future use of the technology, it does not describe the path the technology and the market will follow. The Wildfire scenario reminds us of the General Electric pavilion at the Worlds Fair where sequential generations of a house kitchen were presented on a revolving stage. Although the historical views from the 1930s onward were logical and accurate, an enormous leap of faith was required to understand the "kitchen of the future."

There is a logical path forward, and it will be governed by powerful forces, namely, consumer interest and application value. These are the two same forces that were introduced in Chapter 2 and will be expanded in the following section.

Safe and Effective

The Food and Drug Administration created a two-part formula that has become the acid test for all new ethical pharmaceuticals introduced into the United States. A drug must be both *safe* and *effective*. Drugs that pass the rigorous clinical trials and other tests posed by the FDA are awarded the official stamp of approval and are then permitted access to the market at large. For a major drug, this could mean the chance to earn one to five billion dollars a year.

The parallel to the world of wireless data is quite straightforward. Products must pass two acid tests. First, they must be *convenient* meaning that consumers can readily purchase and use the product, and second, products must be *desirable*. Products that surpass these two tests can achieve large-scale success in the marketplace. The parallel drawn here continues beyond the tests themselves to the arbiter of the test scores. The FDA publishes its standards, and although any particular pharmaceutical firm may find the regulations and criteria difficult or harsh, the regulations apply uniformly. The FDA creates the rules, and they perform the evaluation. All begins and ends with them. The standards body in the case of the wireless data market is not some third party organization disembodied by design from the market process. In the wireless data market, as with other computer software and hardware, the judge and jury of a product's worthiness is carried out by the buyers themselves.

This makes for a fickle court forcing product vendors to contend with an unpublished set of guidelines and a moving set of criteria. In this court, the "best product" may not win in the marketplace because the "longest list of features" is not necessarily how the marketplace will measure a product. Before defining how the wireless court will measure convenience and desirability, a brief example will prove the point.

Lessons Learned

In the late 1980's Microsoft and IBM went head-to-head in a battle over the desktop operating system. At least that is what IBM thought the battle was about and as such, they put forth a strong technical argument as to why they had the best product. OS/2 offered crash-protection, full 32-bit processing, multi-tasking, pipelining, and a long list of other important features. Microsoft seemed to be on the ropes since it didn't counter with any such claims of product superiority. Unfortunately for IBM, the battle was not over features as customers in the PC market knew or cared little about pipelining or 32-bit processing. What they did like were applications. Applications were desirable since they did something whereas an operating system appears to the user as doing very little. Users can balance their checkbook with an application but what could they do with an operating system other than tell it the time of day and the day of the year. So, Microsoft's Windows 3.1 product offered both *convenience* since it led users through the installation process, and made the computer interface easier to grasp while at the same time Microsoft partnered with hundreds of application developers who offered applications that were *desired* by customers. IBM had done poorly on desirability and only modestly well on convenience. The verdict of the PC court is well known.

Wireless Convenience and Desire

As the nametags of these forces imply, the motions of a market are driven by a ying-yang tension between facts and emotion. Convenience is measured in relatively matter-of-fact dimensions of time, level of skill, and distance one has to drive to find a retail outlet. Desire differs considerably because what is desirable may not seem obvious. For example, SMS is extremely popular yet it is unlikely consumers would have demanded such a service if it was described to them. Picture consumers getting excited about having to make 13 keystrokes just to type "hello." Obviously, they are not excited about the dexterity exercises, but they are enthusiastic about communicating when wireless.

This distinguishes one of the lessons learned for mass acceptance of wireless data. Users do not have to rate both convenience and desire high. As the Figure 5.3 shows, desire is more powerful in determining success than convenience. This point is often lost on technology companies, including network providers, as they are expert at technical speeds and feeds, which are more closely aligned with convenience than with desire. Unfortunately, this results in many technologies being sent to market only to be rejected not only in the wireless marketplace but in technology markets in general.

Convenience for wireless data can be viewed in parameter sets such as the one listed on page 103 paired with a starter set of questions. Scores against these questions would determine a product's or service's convenience. The particular questions listed will change depending on the product or service. For example, if the wireless data offer in question is a GPS tracking system for a bicycle, then the question for "weight of the device" would be "does the device add significant weight to the bike as to make it noticeably heavier?"

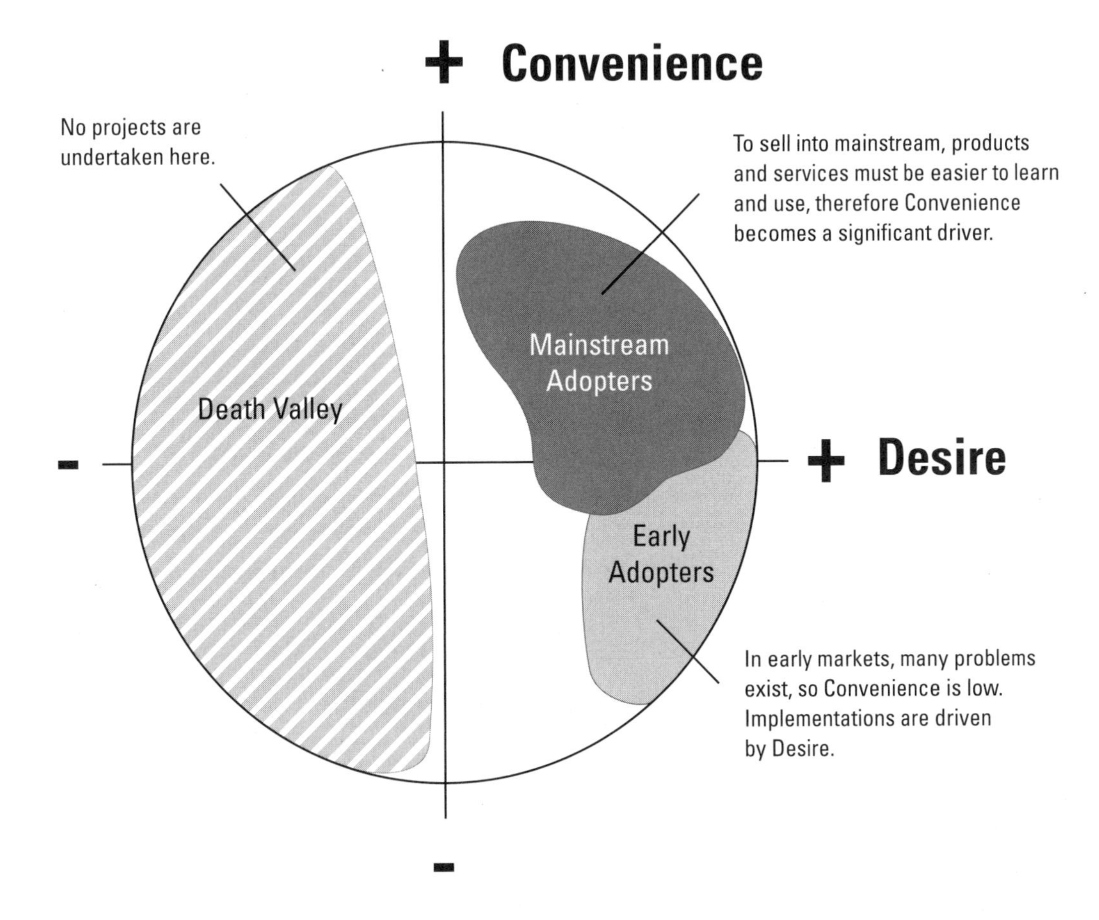

Figure 5.3 Convenience and Desire = Adoption

Convenience Factors

- weight of the device: can a user carry it around all day?
- size of the device: does it fit into a pocket or a purse? Can it easily be worn?
- durability of the device: will it work if dropped?
- battery life: is the battery life at least a few days?
- time to get data: do users get data across the network before they get disinterested?
- time to boot device: does the device boot time take so long as to cause the user to say, "it takes too long – I'll do without the information?"
- screen size and readability (color vs. mono; pixel density): is the screen useful in a variety of ambient conditions? Does the screen provide enough information? Does the screen provide the proper level of density or color to render the content appropriately?
- ease of interconnecting to other devices: is it easy to connect this device with another?
- availability of training: does the user need to be trained on how to use the product?
- availability of the device: is the device easy to find? Does it come assembled or is the user responsible for putting together the pieces?
- input facilities (voice recognition, pen, mouse, graffiti . . .): Is it easy for the user to interact with the product?
- cost: is this a product that is priced to be used by the masses?

The list is quite helpful when looking at market uptake. Failing convenience factors will likely kill a product's chance for large-scale market adoption. Today, for example, users who wanted to reach their office e-mail system while wireless are responsible for buying a PCMCIA card, identifying the wireless carrier that matches the card, identifying and installing the proper software on their PC and on the server, and making changes in their e-mail server. Mass market is all about a simple, out-of-box experience. Currently, most of what is offered in the wireless space is designed for the technically skilled, which by definition, significantly limits the addressable market for wireless data. Compare this with BlackBerry service. A user can purchase the product, the wireless service, and the application all from a single source such as a web site. The user interface is easy enough that most features can be learned without opening the manual, meaning that a user can be sending e-mails 15 minutes after they receive their RIM 950/957 in the mail. The BlackBerry also does well against the other convenience features.

Desirability of a wireless product or service gets directly to the heart of the issue of technology in our lives: to what degree can we live without it? Most technology we encounter during the day requires little from us and therefore the level of desirability that it must deliver is not high. Garage door openers are probably one of the most basic "wireless" devices with a single button to open the garage. Once installed and programmed, there is nothing to do except press the button to raise or lower the door. Therefore, these are quite convenient. How desirable is the garage door opener? Sufficiently desirable that tens of millions of them have been sold. Why? Because Americans are trained to let machines do work for them, and a mechanical garage door has less status.

The Desire for Status

The idea of status is typical of the parameters that characterize desire. They are emotional elements that are softer than the hard-edged list that describes convenience. Desire is peculiarized by emotional terms – it is a want not a need like convenience. Herein lies the power of desire as a way to grow a market. Wants control spending in dramatic ways that needs do not. Needs tend to be planned and as a result, grow in increments. Wants on the other hand, have the potential to explode.

There is no corresponding list of wants that describes desire. It is also not necessary to meet or exceed several items that link to desire. One item is powerful enough to pull a product into the limelight of the market. Having stated there is no list, below are some typical connections to desire: status, power, exclusivity, intelligence, and the elusive "cool factor."

Three Vectors for Growth

No one is so prescient as to didactically describe how the wireless data market will evolve over the next few years, although curiously enough, many have made predictions with scientific certainty attached. The acid tests described in the earlier section are the filters that will determine what products and services will reach large-scale acceptance, however, the tests themselves do not determine what products and services are brought forward. Going back to the FDA analogy, the drug companies are responsible for creating new products. In the case of the wireless data market, software vendors, device manufacturers, and network operators are collectively responsible for bringing forward new products. Instead of looking at any particular products or services, it is more instructive to look at classes of applications.

First and foremost, business-to-employee applications such as pickup and delivery applications have led the way in the wireless data market. These have an ROI and attendant soft business benefits that make them highly desirable. Convenience of these applications have not been high since early devices were bulky and heavy, and wide area wireless networks were slow and had poor coverage. Since an organization could mandate that their employees use the new wireless data system, convenience became a less important factor.

Moving forward into the broader commercial marketplace, convenience will take on a stronger role, and as is the case with disruptive technologies as they approach the mass-market, the technology will have to be responsive to new requirements. Although monochrome displays were sufficient to meet the needs of road warriors or the FedEx delivery driver, mass market devices will need to deliver full color since in the mass market the device will be used to render entertainment.

Three Classes of Wireless Applications

Three classes of wireless applications are coming on strong. Like Shakespeare's man vs. man, man vs. nature, and man vs. god themes, the wireless applications can be segmented along the lines of humans and machines.

▸▸ *The Need for Machines to Talk to Machines – Application to Application*
Wireline applications such as thermostats that already tell air conditioners when to turn on and off will naturally extend to a wirefree model. We already see the early beginnings of this with the Notifact/Aeris energy control systems that use a wireless connection to the Web *(See Chapter 4)*. The systems send performance data to a monitoring center, which uses "smart agent" technology to keep tabs on the health of the HVAC systems. When "out of boundary" performance is detected, or when mandated shut downs are required, performance tuning and switching instructions can be sent automatically over the air. This will be extended very quickly to automobiles where engine performance will be sent to "electronic garages" where smart agents will continuously update records. An oil change can be scheduled automatically based on the odometer reading, or as these agents become more sophisticated, oil changes will be scheduled on the severity of the miles driven (e.g., city driving, which is hard on oil and engines, vs. easier driving highway miles).

▸▸ *The Need for Machines to Talk to Humans*
Telemetry applications will talk to humans, both fixed and mobile. For example, a vending machine inventory check updates a host application, which then advises a delivery driver where to drive and what items to bring. This application set can be extended to a wide variety of known and future applications in the utility, insurance, health, government, and manufacturing industries. Architects are beginning to explore "smart walls" where humans can talk to the wall and have it act as a television, artwork, or interactive white board. Automobiles have been "talking" to us for years, albeit in relatively immature language. Voice-activated commands are likely to be the first common experience society has with interacting with a physical object and having it obey our commands.

▸▸ *The Need for Humans to Talk to Humans*
We have already seen the first successful wave of applications in the human-to-human wireless data space. The application is messaging, and it has taken the form of SMS in Europe and BlackBerry services in the U.S. With billions of messages per day being sent, the application is a breakthrough – simple to learn and use, cheap, and indispensable. As such, messaging is invaluable to the evolutionary process since it is both application and educator of the masses. It will, however, be superceded by new applications that will make

SMS seem primitive in just a few years. Collaborative messaging, a step in this bold new direction, has already arrived. The popularity of instant messaging is quickly growing in the wireless space.

Computing Can Give Insights into Wireless' Future

We have only begun to imagine what applications can be enhanced or created using an anytime, anywhere mentality. Mired in what we know, there is a natural gestation period required for technologists to become facile with features and subtleties of the wireless world. As mentioned in Chapter 2, "needs" and "technology" collaboratively push each other. The first generation of applications, as described in Chapter 4, are useful to enterprises and deliver competitive advantage, market reach, revenue generation, and cost savings. Yet it is fair to say that wireless applications are initiatory and will mature considerably in the next few years. This can be stated quite emphatically given a number of parallels in the recent history of computing:

- *PC applications at first directly mimicked mainframe applications.*

In 1981, when IBM introduced the PC, it entered the computing landscape not as a standalone computer so much as an appendage to the mainframe. Granted it was offered by IBM who dominated the computing industry at that time with its "big iron." However, what kept the PC within the gravity of the mainframe was the lack of pull to move it away. It was easy to see how to make the PC an expensive terminal by using its flexibility to place add-in cards for attaching coax cables and running 3270 emulation software. What wasn't as easy was to know how to use the PC as a PC.

This is a natural trend for new technologies, especially disruptive technologies, as they are often erroneously force-fitted into existing paradigms. Only human creativity envisages new application types for disruptive technology, and this creativity evolves after the new technology is born. With new applications, a broader sense of utility and practicality is created.

Reviewing their growth in popularity, PCs did not flourish in the beginning. IBM talked about them as "mainframe terminals" – as appendages to the core mainstream technology of the day. This is a common strategy: new technology is "named" (and hence belittled) by existing technology. As such, the first PCs were very expensive and not exploited since terminals rely on the server (mainframe) for almost all of their processing (application logic, workflow, database, are all executed on the server). While this centralized processing approach is appropriate for an inexpensive terminal (called "dumb terminals" because they lacked processing capabilities and cost only $400 circa 1982 compared to an IBM PC XT that cost $3,500), a distributed processing model is much more in-line with the rich capabilities PCs have to offer. It took a few years, but spreadsheets and word processing quickly took hold (the parallel in today's wireless data world is messaging) and spawned a generation of new thinking about what a PC could do, both standalone and interconnected with other PCs and servers. This helped create applications such as gaming and animation, which emphasized the rich graphic capabilities of the PC. The natural evolution of wireless applications is for them to find "themselves" and to establish their merit as a new class of applications. As with the PC, there will be merit in standalone wireless applications as well as applications that inter-operate with landline applications.

▸▸ *Wireless devices are evolving from a wired paradigm. It will take time for them to evolve as well to be truly wireless.*

Remember, disruptive technologies do some things better (or different) than their predecessors. In the case of the PC, although its processing power was dwarfed by that of the mainframe, the PC's graphic capabilities were far better. With more colors and greater graphic speed in rendering shapes, applications such as drafting (2D) and solids modeling (3D) moved from mainframes to PCs as the processing power of the PCs increased to make the necessary computations.

Similarly, wireless devices are portable – this is their one significant differentiation that helps them to be disruptive. If they are to reach the mainstream they will need to improve in many ways. Laptops are a perfect example. When introduced, laptops were purchased for the mobile user. The devices were much more expensive than a desktop computer, they had less everything: weight (a good thing); screen size, disk space, memory, and processor speed. Even with all of these drawbacks, laptops moved from their mobile-only market to now be considered as desktop replacements for people who are only occasionally mobile or who want to be able to roam within their building, such as programmers. To be so successful, laptops now rival desktops in all aspects of computing prowess.

Wireless devices will need to make improvements in the following areas:

- *Input.* Either from a human (such as typing, voice input, or mouse clicks) or from a machine point of view (such as engine information being collected from pressure gauges, flow meters, and crank speed), input devices are essential. Today's input mechanisms, such as phone keypads for text input, are generally accommodations for the user and not a satisfying user experience. Voice recognition is a major step forward that will greatly increase the acceptance of mobile devices. Gesture gloves may be another, where the user moves its hand in space and this motion is interpreted such as in moving a mouse.

- *Output.* This typically takes the form of a screen, but it is any output that can be identified. For humans, speakers in our phones are a form of output. Similarly, graphical displays (vision), vibration, temperature changes, air motion, changes in smell . . . are all valid forms of output. Output needs to move away from the hard, small, monochrome screens to information that is portable, stretching in the direction of heads-up displays that are holographic.

- *Computing.* This is where the ones are added to the zeros to determine bank balances or the length of a line in a 2D drawing. Virtually all wireless devices incorporate computing elements to exchange, store, and process information. Processing speeds need to dramatically improve if there is a demand for video processing or other entertainment applications.

- *Storage.* This is where applications, data, and logic are stored. On a PC, storage is on the hard drive. On a PDA, it could be flash memory (solid state vs. the spinning disk in a PC). It will be mandatory to grow the current storage by a factor of 100 to 10,000 to be able to easily store a DVD movie or other rich content.

- *Network.* Improvements will need to be made in both the personal area network as well as the wide area network. Locally, Bluetooth or other technologies can address the convenience issues of connecting a PDA with a keyboard. The singular biggest challenge is increasing the available bandwidth to a wireless user. Today's effective bandwidth of 8 to 20 Kbps needs to grow by a factor of at least 10 to begin to deliver rich content such as streaming audio or video. Of all the technological challenges, this seems the most daunting.

Factors for Mass Adoption: Wireless Web and Bluetooth?

Up until now, wireless data adoption on the consumer scale has been rocky. Carriers, analysts, and specialists in wireless content have equated the "untethered lifestyle" – mistakenly or not – with something along the lines of "Wireless Internet," i.e., extending the Internet's reach via voice-centric wireless networks to the average consumer. This has had mixed results. In early incarnations, smart phones equipped with small LCDs, circuit or packet data capability, and a Web browser allowed consumers to establish minimal, narrowband connectedness with a limited number of Web sites (and just tiny amounts of content, most of it poorly displayed). Using the acid tests, this fails miserably in the dimension of "convenience" and scores poorly in "desire." In ensuing generations, specialty wireless Web services such as Omnisky beefed up wireless Internet content, enabling better accessibility to richer, more capable wireless sites, giving an advanced generation of PDAs and palmtops access to content optimized for mobile users.

Today, Personal Digital Assistants with larger screens and more memory and CPU fire power can link to selected wireless Web services and, by extension, wireless intranets, the secure corporate version of Internets using Internet Protocol (IP) connectivity. When properly configured, these connections enable access to enterprise applications and files, as well as consumer-driven services, such as traffic reports and GPS directions, snippets of news, entertainment, sports, horoscopes, e-mail and short messages (on the consumer side), banking, brokerage, directories of local businesses, and some transactional sites to buy and sell things. In terms of convenience and desire, these offerings begin to break through the minimum thresholds and, therefore, hold market promise.

Short Messaging Service (SMS) is a booming business in Europe with more than 40 billion messages exchanged on wireless phones each month. Wireless instant messaging is poised to follow this same trend to wide-scale usage. Teenagers are prolifically attached to wireless data, gaming, messaging, and dating on-line while chatting, both through data and voice, on GSM, *General Packet Radio Service* (GPRS) and i-Mode phones in Europe and Japan. Similarly, wireless infrastructure providers now promise a revolution in short-range wireless communication with Bluetooth; a so-called "cableless" connection standard that eliminates twisted wires. Bluetooth enables up to 8 wireless devices to "talk to each other" at short ranges over the air (up to 30 meters); it's actually a specification for a tiny radio transceiver substituting for the infrared port commonly found in most computers today.

Will Our Devices Talk to Each Other?

If widely adopted, Bluetooth could enable short-range piconets to flourish at home, in the office, in airports, and factories. These connections will link wireless phones and laptops, PDAs, pagers, home-grown wireless LANs, and industrial/barcoding terminals and scanners, allowing them to communicate and synchronize data automatically — a very useful function given the millions of today's devices snarled together with a confusing array of unique cables and plugs designed to infuriate any human who comes in contact with them. The goal of Bluetooth is to make wireless computing ultimately super convenient.

Bluetooth represents a fluid way of establishing cheap connectivity in mobile environments. For example, although the Bluetooth specification lacks the network management capabilities of a full-blown wireless LAN, it may, when coupled with the Web browsing, boost consumer wireless usage. Bluetooth could help make wireless communication cheaper and broader in scope, larger in geographic dimensions, and "plug and play." When coupled with Web phones and PDAs featuring WAP (Wireless Access Protocol) or other wireless browsers, Bluetooth could, in theory, enable millions of wireless consumers to exchange information and perform electronic transactions such as hotel registration, point-of-sale transactions, data exchange with infotainment kiosks, airline check in, and other services.

Why are Bluetooth and the Wireless Internet important together? Some industry analysts still believe they were stepping stones to wireless ubiquity and the elusive American "mass market." Witness NTT DoCoMo's i-Mode wireless Internet success story, the only true commercial service on earth to enjoy explosive mass-market success. From launch in February 1999, i-Mode, which provides Japanese consumers with mobile services such as telephone directories, restaurant and ticket reservations, on-line banking, and automatic, always on e-mail grew from 220,000 subscribers after just three months to 21 million within two years. Nearly 100 companies have linked their i-Mode Web pages with DoCoMo's portal IP Web site; 550 companies and individuals had independently launched i-Mode websites by April 2001 – and this with a service offering no more than 9.6 Kbps data rates! Some reports indicate that as many as 31,000 unofficial Web sites are now accessible through i-Mode, which does not require a special Web browser or translation language, but is based on a reduced set of the HTML global Internet standard, dubbed compact HTML or cHTML. U.S. carriers and content providers are struggling to replicate the Japanese success story. Explained Kei-ichi Enoki, director of DoCoMo's Gateway Business Department: "A prominent reason for i-Mode's success is its simplicity," he said. "Both users and information providers view i-Mode as a promising method of communication offering tremendous convenience."

> **THE WIRELESS EXTENSION OF THE INTERNET IS A NATURAL PHENOMENON THAT WILL HAPPEN**

When Consumers Talk

The outlook for wireless data usage in both the mass market and business is still enormously bright – provided consumers and mobile professionals *open their mouths and loudly say what they want.* In the wireless data market, there is an awkward dance of partners who have yet to discover each other. Carriers, for all of their inherent and deep understanding of consumer and business needs in the area of voice, have yet to hear the wireless data music associated with these same buyers. They seem too busy stepping on their customers' toes to learn what their future wireless data partners need and want. Although carriers do quite a bit of research in their typical lines of business (pay phones, long distance, and voice service), they have not undertaken sufficient research as yet in the area of consumer or enterprise wireless data. Most work is strongly biased towards technology, e.g., "would you buy a DSL connection to the Internet if it were made available in your neighborhood?" Or, "would you be willing to pay $49.95 a month for a wireless connection to the Internet on your laptop?" Most wireless operators and infrastructure providers are struggling to understand consumer tastes and preferences for wireless data at the same time consumers are trying to define it for themselves. It seems most likely, then, that wireless data will continue to progress as described in Chapter 2, with a collaborative force between "market needs" and "technological do-ability."

TEENS are one of the great untapped markets for wireless data in the U.S.

The Forgotten Markets for Wireless Data

Here are some applications for wireless data that are frequently discounted or overlooked.

- *Teenagers – the "Gen Y" crowd.* The Internet Generation, "Gen Y . . . has been exposed to more information during their youth than any other age group – making them a superb target for wireless data. They are more eager to use information, have high expectations about having it "anywhere and anytime," and have no technological barrier that tells them they can't have information sent to them while they are driving at highway speeds. Teens are one of the great untapped markets for wireless data around the world. Teens currently pump $160 billion annually into the global economy. Their preferences for video games, paging, chatting, sending messages, downloading books, music, and movies make them naturals for wireless data. Over the next few years, as device providers integrate such functions as MP3, multiplayer games, e-mail, instant messaging, and buddy lists – wireless data will capture the teen market. As "Gen Y" marketing maven Dave Bell (CEO of Chasma Inc.) suggests, "What teen will need a Nintendo Gameboy if a cell phone can play games of equal quality – and more?" Big qualifier: Cost to parents.

- *Gameplayers, all ages:* Internet gaming via mobile phones will reach 200 million people in Western Europe and the United States by 2005, representing a $6 billion industry, says Satoshi Nakajima, Founder and President/CEO of UIEvolution. Gaming scores high on the desire scale. The challenge is to make the human interaction with the game satisfactory. This is already being done in Japan with many games on the i-Mode network.

- *Lost Parents, EMS workers, Real Estate Agents:* Dashtop wireless data systems incorporating GPS directions, audio/speech recognition support, and mapping will be packaged with the new generation of cars and trucks over the next few years. Like air conditioning, fuel injection, and air bags, dashtop wireless systems will first appear in expensive vehicles and move into the mass market. Unlike air conditioning, fuel injection, and air bags, dashtop wireless will be used in hundreds of different ways. The applications for this technology will include emergency care (911, personalized medical instructions, "I'm lost, give me directions"), entertainment (games, personal digital radio), and business applications ("locate the next house for sale that meets the following criteria", "send this photo and claim form to the home office"). Like the "Wildfire" scenario earlier in this chapter, the automobile will be outfitted with a wireless world that will connect the driver and passengers to their home, their business world, and their personal lives.

- *Consumers hungry for local bargains.* Wireless Internet browsing is expensive, inefficient, and boring in its current instantiation. However, local merchants cooperating with carriers and mobile virtual network operators (MVNOs) can create compelling mobile e-commerce transactions by using phones and PDAs as a vehicle for local advertising and merchandising.

- *Instant Messaging.* Instant Messaging (IM) has tens of millions of users on the wired side of the fence with users accessing AOL and other popular IM engines. Given the interest in always being connected, IM makes a natural portable application. Not only does it work by itself, but it can also be used as a vehicle for many different types of applications such as m-commerce. Consider the situation where an individual joins different buddy lists, such as those offered by their favorite stores. Then information such as new products or special pricing could be shared with the entire community at one time.

One way consumers today can fight boring wireless data is by refusing to accept and pay for unimaginative, expensive, and low-value services – the kind we have seen in "first gen" North American wireless Internet carrier offerings thus far. Consumers need to shout out what they need and want; carriers and content providers need to listen.

Most likely, a whole new generation of device and application specialists, mobile virtual network operators (people who will "run" the wireless data networks for the carriers), consultants, and consumer advocates will be required to implement compelling consumer services. Alternately, wireless users can propose ideas directly to content developers, middleware providers, consultants, and integrators (most likely through their enterprise IT structure). Adult users can send their teenagers out to work for wireless games manufacturers – because chatting, messaging, gaming, and entertaining are indeed some of the truly *proven*, giddy, and money-making applications that promote consumer mobile data usage today.

What is clear is Wildfire is not as close as Orange envisages, yet even if Wildfire is not commercially available until 2010, there are an enormous number of applications that can be enjoyed by the consumer starting today. The PC era has taught many lessons. As creative thinkers apply themselves to the wireless data space, the wireless equivalent of Lotus 1-2-3 and Flight Simulator will emerge followed by a horde of new ideas yet imagined. As a society, we didn't say in 1985, "gee, this is only somewhat interesting, so let's wait a decade before buying a PC". Similarly, we are standing at the time line equivalent of 1985 for the wireless consumer market. It will be a fun and unpredictable period of growth ahead.

Tips for Mobile Operators

Voice and speech recognition could be the key to unlocking the Mobile Internet. Users will gravitate to a device if it is easily understood and simple to use. VoiceXML, a new standard essential to making Web content accessible via voice, and Speech Objects, which significantly speeds up the creation of speech-activated applications are technologies worthy of investigation. The Voice Browser Working Group (W3C) is working to expand access to the Web to enable people to interact with Websites via *spoken commands.* Allied Business Intelligence claims that 11 million voice portal users in 2001 will grow to over 56 million by year end 2005 in North America. There will be approximately 2,000 voice sites by 2001, reaching 250,000 sites by the end of 2005.

The unpredictable – and quixotic – applications and markets will score the "killer app" win for the Mobile Internet. Half of the U.S. youth market will eventually own a wireless phone, predicts Cahners-Instat. Three quarters of all teenagers will use one. Most likely, developers will stumble on highly focused applications that win a wide allegiance through a mix of killer marketing and odd niche preferences. For example, the "Gen Y"-teenage market has great potential with targeted, very affordable data applications including "dating profiles," on-line chat, and instant messaging, also shopping, Napster or equivalent, MPEG 4 video, cartoons, and interactive voice/data, among other applications. Another enabler: handheld PCs/combined with prepaid voice calling will drop down to the price of a pager – and carriers will be virtually "giving away" millions of megabytes of bandwidth for the glitzier "Dick Tracy wristwatch" razzmatazz.

Moreover, virtually all carriers will migrate – albeit at differing speeds – toward a data-capable high-bandwidth solution that will support a minimum of ISDN and higher (128 Kbps and up) wireless data rates. Both voice and data services must be bundled and priced in consumer-friendly ways – and possibly sold and integrated with wireline services (e.g., selling X number of minutes per month, regardless of the device, wireless or wireline; selling Y number of packets per month, regardless of the data pipe). Today, we are seeing the first build-outs of so-called "2.5 G" networks – notably CDMA 1XRTT, a non-contention based, and packet format standard operating at 144 Kbps. This is a stepping stone to the North American standard for 3G – CDMA 2000 – that should interoperate (it is hoped) with the European international standard for wideband CDMA, also known as IMT-2000.

Most likely, data applications (such as those in the bountiful teenage market) requiring slickness and large chunks of bandwidth will drive the deployments of wideband services, not the other way around. Rather than allowing wireless carriers to run the show, we will see a new breed of content providers and applications developers who work for *Mobile Virtual Network Operators* (MVNO). These operators could bring imagination and the power of their branding to wireless data by roping in a suite of already loyal customers (an example could be AOL/Time Warner or Virgin). The MVNO already has an established brand and brings a community of devoted customers.

In short, the “wirefree” and intelligent home lifestyle will eventually become a part of our modern existence. Most of the faltering and inconvenient technological solutions wireless operators have hit upon thus far will be replaced by an “environmental” model for wireless that is, as Orange puts it, “simple, intuitive, and human.”

6 MIDDLEWARE IS THE HUB OF WIRELESS COMPUTING

OFTEN TECHNOLOGY BUYERS WATCH IMMATURE TECHNOLOGY develop much like an observer who muses over a passing parade. Each new element in the parade is interesting, the music is lively, and there is a general sense of fun and entertainment. After a while, the onlooker becomes somewhat numb to each new band or float, and after some time, the crowd disperses and the parade is over. The bands and floats in the parade are the products and vendors of a new technology with each vendor under pressure to be identifiable from those in front and behind them in the parade. The music is the trade press and the marketing excitement made around the vendors and their technology. Recently there were also sideshow pitchmen in the form of investment bankers who were touting inflated values for the companies they backed, adding a sense of heightened excitement to the festivities. The parade over the past few years has been rare and exotic.

Yet, parades are transient – they exist only for the moment and for their entertainment value. Vendors and their products need to do much more than amuse (and confuse) the buying IT and network management "public" since companies cannot survive without on-going sales. There must be permanence to the companies in the minds of the buyer. There must also be a perceived value in their products and services to attract their attention. The vendors in the parade have to move from the Fair Grounds to Main Street, establish storefronts, and operate enduring businesses. In essence, they must convince buyers that now is the time to get started and to purchase and use these exciting products and services.

The wireless parade is similar to the parades before it, such as the minicomputer parade in the late 1970s, the relational database parade of the late 1980s, and the enterprise integration platform parade of the late 1990s. As with these other technologies, wireless has reached that critical point of delivering measurable business value. In earlier chapters, we sampled a variety of companies who have jumped in and got started. The taxonomy of business applications is large and spans multiple industries including aerospace, transportation, insurance, utilities, banking, brokerage, real estate, manufacturing, retail, pharmaceutical, health care, and government.

So, how far is it from the Fair Grounds to Main Street and how do you know the vendors have arrived and set up shop? This is actually quite easy, since the first few vendors to arrive are heralded with fanfare. The trade press provides headlines for unique and challenging breakthroughs akin to the widespread coverage of landing on the moon or the acceptance of Netscape's browser or Palm's PDA. FedEx's package tracking or the first insurance company to provide real-time wireless field claims processing received a hero's welcome by primary technology publications as well as vertically oriented, industry magazines. For wireless, this occurred years ago.

Reaching Main Street

It is more difficult to see how and when Main Street develops for it has to mature to serve the broader and larger market beyond the early adopters. In Chapter 5 we talked about the two drivers of wireless data: *convenience* and *desire*. In the industries listed previously, the early wireless applications were desirable because they had a hard dollar return-on-investment. Convenience in early markets is less a determining factor for the early adopters. However, as time progresses, mainstream buyers will demand greater convenience where the business improvement can be obtained with less and less difficulty (See Figure 5.3). Although FedEx installed its own private radio network, as did IBM in the 1980s, the vast majority of corporations today must leverage existing networks. Likewise, in some early installations, unique devices were designed and built to exacting specifications to meet the distinctive needs of the early customer. Today, companies choose to use more standard networks, hardware, and operating systems to reduce their maintenance loads and the staff skills required to build and deploy wireless applications.

Abstraction Simplifies Technology

A natural progression must occur in a market if the vendors are to be in position to service the broader marketplace and to achieve the level of convenience necessary to make the technology usable by more technologists. The technique used to simplify technology adoption is *abstraction*, and it is a trend we have observed in technology as it shifts phases from early adopters to early mainstream. New technologies that lack an abstraction layer languish until an abstraction layer is offered to the market. The abstraction layer hides the complexities that lie underneath, presenting a more pleasant, and easier to work with, view of the technology. Beyond early adopters, who have the wherewithal to launch their own "mission to the moon," new technologies cannot get off the ground until the abstraction layer is made available. The abstracted layer, as shown in Figure 6.1, provides three quintessential features.

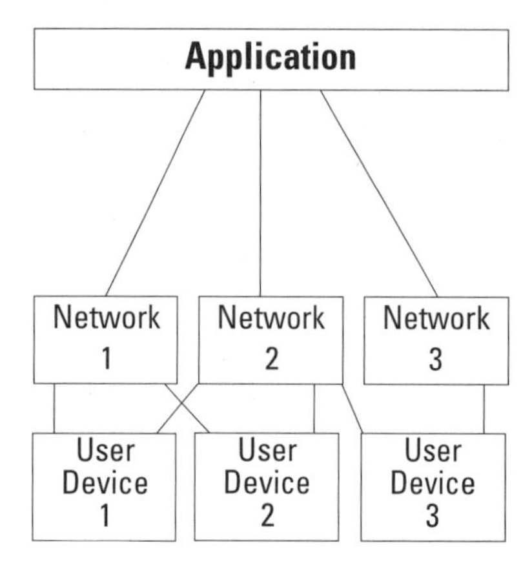
Application
Network 1
Network 2
Network 3
User Device 1
User Device 2
User Device 3
Application developer is highly involved in networks and user devices. Changes in either causes a direct change to the application.

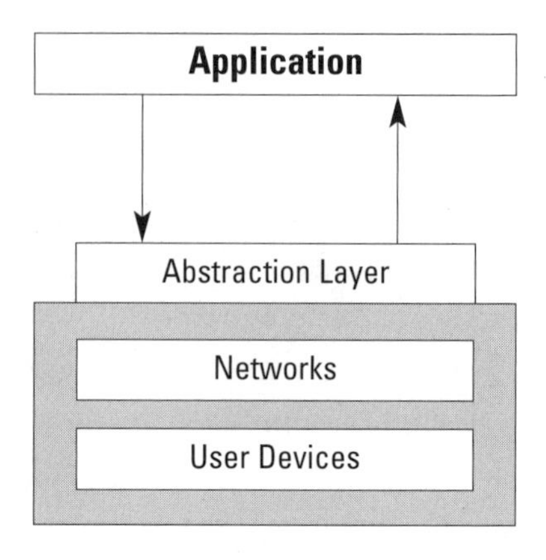
Application
Abstraction Layer
Networks
User Devices
Application developer is insulated from networks and devices and how these change over time.

Figure 6.1 Three Quintessential Features of the Abstracted Layer

Three Quintessential Features

1. *Familiar interface.* Application developers need an easy way to interface with a new technology. In the case of relational databases in the mid-1980's the abstracted layer was an application-programming interface utilizing *structured query language* (SQL). By using this new standard, developers could start to build applications that stored and retrieved information in and from a relational database.

Application programming interfaces (APIs) are a common method of abstracting function from a technology and presenting the developer a set of "knobs and switches" to use. Successful APIs are designed to reduce the learning curve and to provide both development and runtime support. In the case of wireless data, the abstracted layer must provide the application developer an easy way to send and receive data over a wireless network and to handle the variety of new functions needed to work with wireless networks and downstream wireless devices. For example, in a wired application, developers can add security and network management to their application by writing to established APIs. Similar APIs need to be available to the developer so that the wireless application can have comparable richness in function.

2. *Insulation.* The technology involved in moving data across a wireless network is complex and not of great interest to the application developer. Multiplying this complexity by the number of different network types makes the task even less attractive. In the ideal world, the abstracted layer completely shields the developer from having to know anything about the details of the layers beneath. Like an airline passenger who doesn't want to know how the plane works or how air traffic control routes planes and schedules landings, the developer prefers to be relatively ignorant of how data is packetized or how messages are queued for users when they are not in coverage.

Insulating people from technology is not unique to the wireless world. To the contrary, this is exactly what must be done to make a technology more broadly accepted. Application developers know very little about how a relational database system actually works. They have an understanding that data is stored in rows and columns, however, this is a logical view of how the data is stored – not truly how the relational data system stores data on disk or in memory. The "rows and columns" analogy is a useful one since it is easy to understand and manipulate by the application developer. The analogy also serves as a metaphor so that database functions, such as inner and outer table joins, are almost intuitive as to what they might do.

In reality, insulation provided by the abstraction layer is not one hundred percent, meaning that developers still need to learn something about the underlying technology of moving data across wireless networks just like developers do have to learn some basic rules of how to work with relational databases.

3. *Future proof.* Technology evolves continuously. Applications, which are built on top of a number of technologies, must work, day in and day out. PC users fully expect that when a new microprocessor is shipped that the current version of their applications will work. In fact, their expectation is that the application will run faster since the microprocessor is faster. To make this happen, an abstraction layer known as the operating system future proofs the application by working with the microprocessor manufacturer. Future proofing is a mandatory requirement yet difficult to achieve if application developers have to provide this themselves.

Consider the case of supporting a new network or new wireless device as they continue to emerge. It would be impossible for any application developer to work with all of these vendors in advance so that when the new technology hits the market, the application will be ready to work with it. The onus for insuring that new technologies can be incorporated into an application is the responsibility of the abstraction layer vendor.

Abstraction Layer = Middleware

The abstracted layer is generally given a technology-based name such as application development platform or the generic catchall phrase of middleware. Middleware has been used to describe the abstraction layer performing functions such as transaction processing, security, database, and session management. As its name implies, it is software that is in the middle, sandwiched between the application and the underlying technologies (See Figure 6.2).

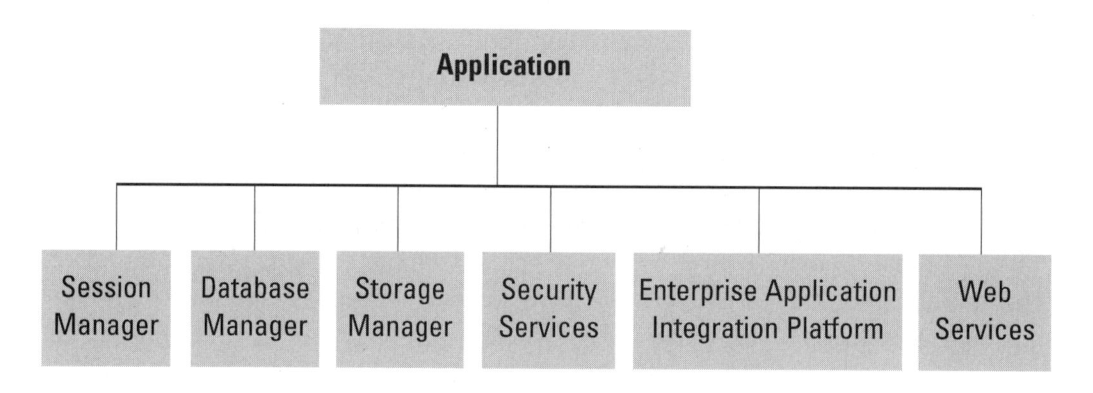
Application
Session Manager
Database Manager
Storage Manager
Security Services
Enterprise Application Integration Platform
Web Services
Applications moved from vertical integration to use of standard services beginning in the 1970's. This reliance on abstracted functions is fundamental to the success of application design today.

Figure 6.2 Applications Rely on Abstracted Services

Middleware touches every aspect of a wireless application and, as such, is a critical element of the total solution, albeit less visible than the pieces it touches. Middleware interacts with wireless devices (phones, PDAs, laptops, etc.), the wireless networks (variety of technologies and multiple generations), IT infrastructure (databases, web content, XML, SOAP, application servers), and the application's logic.

Now, When Don't You Need Middleware?

While middleware is a critical element to most wireless applications, there are some instances where the use of middleware isn't warranted. For example, you don't need middleware when you are undertaking a single application that will not be modified in the future or upgraded in order to use new networks or add networks. In other words, when there is complete stability. This could be the case when the application is well defined and limited in scope and geography by some hard business issues, like a bank that operates in a small geographical region and does not have a business plan to expand. In this example, the banking IT department

As its name implies, it is software that is in the middle, sandwiched between the application and the underlying technologies

may want to develop an application to send depositors a mobile message when their checking accounts reach below a certain limit, say $250. This might require only a simple SMS message, which is a function directly supported in many wireless networks. In this limited example middleware is unnecessary since the network provides this as a base function. Another example sometimes occurs in the telemetry market where machines such as road signs are being updated with new traffic information. In this case, once the system is established it is likely to be quite stable and not have changes over time.

When You DO Need Middleware

If a large carrier provides a complete mobile application – for example, a bundled offering supporting wireless stock trading or brokerage access to mobile employees and customers – generally the richness of this application is best supported through a middleware layer. The use of middleware will ease the initial application development, however, its largest benefit will be when application improvements are made. The middleware will shorten the time it takes to make those changes and require less skill on behalf of the application developer.

In general, applications rarely resemble the limited function applications described for the regional bank or the road signage telemetry application. Most applications have a life of their own, and as a direct result of business and technology changes, applications are modified on an ongoing basis. Between sixty and eighty percent of an application development budget is for maintenance, leaving only twenty to forty percent of the budget for new application development. This large apportionment to maintenance provides direct evidence of the amount of continual effort it takes to keep an application vital. As application developers perform these upgrades and improvements, they prefer to focus on the business logic to insure the proper outcomes occur. It is critical that bank statements properly reflect the amount of money that the customer instructed the application to move from her savings account to her checking account. Making sure that the user, who is now accessing the funds transfer application with her Palm VII PDA require effort on behalf of the developer. Unfortunately, application developers are expert in business logic but unschooled and unskilled in data presentation on the wide array of wireless devices, each of which has its own form factor and features.

A quick example illustrates this issue. A Palm PDA has a larger screen than a smart phone. It also provides a richer user interface since it supports drop down list boxes that reduce the amount of space required to display information as compared to a simple list of items that is available on the phone. To build a user interface for the funds transfer application, the application developer will have to build separate and distinctly different user interfaces for these two users. As a result, the Palm interface will take fewer screens and will have room for some text help on the screen. The smart phone interface will span multiple screens in order to capture the proper amount of information such as the customer's account number, security code, and amount to be transferred. This problem is exacerbated every time a new device is manufactured, since it will require another customized user interface. Without technological assistance, the developer suffers from an intractable disorder.

To avoid spending the bulk of their effort on coding to the expanding array of nuances of each new device, (and of the variety of networks not mentioned in this example) developers can offload this work to a wireless platform or middleware vendor. Since this is a repeatable problem, it can be automated to a large degree by the wireless middleware, saving the developer considerable time and effort. Middleware vendors codify templates and means to deal with this multiplicity of devices and networks, thus greatly simplifying the task for the application developer. This is part of the value of using middleware in the design of a wireless data application. Middleware, therefore, is generally needed and advisable when designing and building wireless applications.

The Functions of Wireless Middleware

Whereas middleware used to refer principally to low-level translation functions in the protocol stack, today it enables applications development in an easy way with more high level functions. Consequently, middleware provides a framework in which to build effective applications – anything from bar code reading to messaging, calendar, office automation applications, or access to corporate directories, ERP systems, and the Internet.

Wireless middleware handles all aspects of wireless communication from security, to filtering, to compression, to dynamically rendering data for different devices. This allows application developers to concentrate on their core business and not have to worry about understanding wireless or developing applications for each new device or network as they emerge. The functions of wireless middleware can be broken down into three categories.

Networks

Wireless middleware supports a wide range of networks, including IP and proprietary protocols. (Note: Networks will be discussed in detail in the next chapter.) To improve the performance of wireless communication, middleware optimizes the data that is sent over the air and minimizes overhead to speed throughput and reduce airtime expenses. To overcome coverage holes and enable effective wireless communications, middleware can also provide store-and-forward message queuing that stores messages when out of coverage and pushes these messages when coverage is restored.

Wireless middleware can also enhance usability through features such as auto-connect, auto-reconnect, and auto-disconnect – features that manage the wireless connection so that the user does not have to get involved. For example, if during a data transfer the user drove out-of-coverage, the user would be required to notice that the transfer did not complete, step through the process to re-establish a network connection, and monitor the success of the transmission. This places the burden on the user, which is a poor application design because it wastes the user's valuable time managing a telecommunications link. A better design is where the middleware monitors the connection, senses the break in service, waits for the presence of network coverage at some point in the future, auto-reconnects to the network, picks up the transmission where it left off, and then auto-disconnects the device upon confirming the delivery of the data.

Middleware also handles harsh network conditions. It can automatically adjust to fluctuating coverage conditions, which are quite common in wide area wireless networks. By relaxing timers and slowing down and speeding up when appropriate, middleware can actually "stretch" coverage holes to minimize connection loss.

Devices

Middleware understands the form factor, operating system, and wireless modems for different devices and provides a standard interface for deploying applications across multiple devices. Middleware can also dynamically render a wide array of content sources to a broad range of devices. This is critical given that wireless devices are experiencing a period of product divergence where new vendors are entering the market with a continuous stream of differentiated products.

These devices share little in the bounds of their shape, size, operating system, chip set, memory, and user input mechanism. This rampant barrage of intellectual capital is diametric to the world of the desktop PC where conformance to the Wintel standard makes PCs from different vendors "plug compatible" with each other. The world of wireless devices will continue to attract the best and brightest product designers in the computer and entertainment industries. Their collective creativity will drive product innovation for the next several years. Although devices will be covered in detail in the next chapter, it is sufficient at this point to assert that this multifarious world of devices confounds the application developer. Luckily, middleware is the antidote to this pernicious problem.

Server Resource Connectivity

Within the computing infrastructure exist electronic resources in the forms of applications, data, web pages, data schema, and directories. These pieces are interconnected to deliver tangible benefits to a corporation. A customer database can be linked to a web page, which is part of a new application to provide a clothing catalog's customers the ability to order clothes on the web. The same database can be used by several other applications, such as billing and customer profiling, since the customer database contains basic data about customers.

This re-purposing or re-use of resources makes information technology very powerful. Upon inspection it also proves that new technologies tend to be incremental to existing technologies, adding value to what currently exists. When the web blasted onto the scene in the mid-1990s, web front-ends were placed in front of existing applications (dubbed legacy applications) making the web incrementally valuable. It did not, as many technology zealots predicted, replace all other forms of computing or communication, rather it proved to be another useful channel for these activities.

Society experienced a dramatic example of this re-use during the latter half of the 1990s. In that short period of time, practically every company on the planet provided their customers with web-based information and applications. This tremendous spurt in application growth was made possible in part by the re-use of existing non-web server resources. In many cases, the new web applications were created by placing a web graphical user interface in front of the old non-web application.

Likewise for wireless applications, many current non-wireless applications can be re-used or redeployed with wireless devices by providing a wireless adaptation. Middleware integrates tightly with existing enterprise applications, allowing companies to leverage existing resources, rather than having to build all new applications that need to be separately maintained. Many middleware providers support delivering web content to browser-based devices. Ideally middleware should support access to applications and web content from browsers and devices with intelligent client applications.

Wireless is Added to the Mix

Wireless data has already proven itself as a valuable incremental technology, just as the web proved several years ago. When businesses implement this technology, they keep their other technology investments, adding wireless data to mix. Thinking back to disruptive technologies, this is expected since the nature of a disruptive technology is to provide some unique function. Laptops provided mobility and were additive to the PCs that were on the desktop. Over time, a disruptive technology's path curves it from the periphery directly to the mainstream center as laptops are now challenging desktops in many companies. Similarly, wireless data has entered corporations as a value-added component to the infrastructure, such as in a field service application that utilizes the existing customer information file for addresses when it dispatches the field service representative. After significant improvements in networks and devices, wireless data's destiny as a disruptive technology is to replace landline connectivity, unplugging all devices from network cables.

In general, middleware should make building a wireless application like building a wireline application. The application developer should be able to work in familiar environments and extend these environments to wireless device users. Although not an exhaustive list, middleware should integrate with de facto enterprise computing standards such as J2EE, SOAP, XML, ODBC, SQL, and HTTP(s).

Middleware also handles many functions that enable safe and effective interaction with these server resources such as security, personalization, notifications or alerts, and more. For example, users may be ecstatic that they can access customer data, but if they have to constantly check to see if that data has been updated, the experience soon deteriorates. It would be much easier if the user could be notified when there is an update to her customers' information. And, she may wish to receive this alert via a short message on her smart phone and then, depending on the type of information, access it on her PDA. Personalization understands her needs and desires and delivers information appropriately.

How Do You Choose Wireless Middleware?

In Chapter Five we asserted that it takes two elements, *desire and convenience,* to create an atmosphere within a corporation that is sufficient to motivate them to deploy a wireless application. The specific quantities of each factor differ between companies however. One lesson that has been proven in hundreds of installations around the world is that the desire to change a business process with wireless data is quintessential. Convenience, especially in this early market of wireless data is not expected, nor a primary driver, and it takes a second seat to desire in propelling businesses to undertake a wireless data project.

Desire comes from a business executive who wants to modify a business process with the expectation that it will reduce costs, improve customer retention or improve profit. The technology that is necessary to improve the business process is interesting only as it applies to the business problem. As a disembodied science, technology is flat and dull and remains the domain of the information technology department. It gains life only when it can be applied to solve a business issue, therefore, the starting point for the selection of wireless middleware begins with the business issue and not with the technology itself.

This may seem somewhat counter-intuitive, yet it is perfectly grounded in good information technology practices where business requirements definitions are done as a prologue to choosing the technology. Too many times, technology has been chosen first only to suffer from "if I have a hammer, everything must be a nail" syndrome. This was the case with Wireless Application Protocol (WAP), which over the past two years was uniformly applied to every wireless need only to fail miserably since it was not designed to handle the variety of needs of wireless users. Leading with technology generally results in fruitless projects and frustrated users. This holds true for wireless and non-wireless technologies. As an important software platform for the enterprise, wireless middleware must be chosen by first defining the business problem.

Defining the business problem is a well-documented subject unto itself, however several key factors enter specifically into the realm of wireless data. These specific issues will be treated in this chapter. The reader is encouraged to explore information technology best practices guides for a more general discussion and understanding.

Defining Your Needs

The questions below are presented to help the reader to define which type of wireless middleware will best suit their needs. There are three types of middleware function from which to choose (see Table 6.1 below). These three types can be blended, as they are not mutually exclusive. By answering the questions below, the middleware type(s) required to meet the needs of the business will become clear.

Table 6.1 Three Wireless Middleware Segments

Middleware Type	Function Delivered	Applicability
Transcoding	▸▸ Translates HTML to WML or HDML	▸▸ Supports a user's need to browse while mobile
	▸▸ Makes it easier to re-use web content on wireless devices which don't "speak" HTML	▸▸ Users must have a network present to be able to access the content or application
Message Queuing	▸▸ Guarantees message delivery over a wireless network by using a store and forward technique	▸▸ Supports a distributed computing environment so that applications can continue to function even if there is no network coverage
	▸▸ Handles the complexities and weaknesses of wireless networks such as out of coverage situations	▸▸ Supports complex wireless applications

Table 6.1 continued

Message Queuing ..continued	▸▸ Supports real-time (interactive) data sharing either pushed from server to device or from the device to the server	▸▸ Requires programming
Synchronization	▸▸ Synchronizes mobile databases with server databases	▸▸ Supports a user's need to ensure that the data they are using is the most current
	▸▸ Good for updating information that does not have time criticality	▸▸ Typically used as a batch method to share data with mobile users

Who is the User?

In most areas of business where technology is applied, the first question rarely is the user's employment status, yet this is an insightful question in the area of wireless data technology. Although, it is not probably obvious at this point, the answer to this particular question has a direct bearing on the downstream technological implementation. To provide insight into why this is such a deterministic question, we will jump to an abbreviated discussion of mobile devices.

Finding the Right Device

Intelligent devices are those that can have a business application loaded onto them and can process this application without the need for a wireless network. Examples are PDAs, PocketPCs, and laptops. Intelligent devices can handle robust applications and rich user interaction. The downside of these devices is they are complex computers requiring skilled staff to install and maintain applications. On the other end of the device spectrum are smart phones that have very limited device capacity and rely on server applications. Phones are lightweight and useful, do not require skilled staff, but can only provide the user with an application when there is a network present.

If the user population of the proposed wireless application is comprised of non-employees, then it is likely that the application will have to be limited to phones and other browser-based devices. By limiting it to devices that rely on server side computing resources, the implementation is simplified. This is critical when dealing with consumers.

If the user population will be entirely employees then the device can be of either type. Employees of a company such as truck drivers, field repair technicians, or insurance claims processors have an obligation to use the technology that is given to them. Intelligent devices can handle complex applications and rich data content, which are likely to be necessary to accomplish their business tasks, making an intelligent client the better choice for them.

The device type chosen and the mode of operation required have a direct effect on the type of wireless platform support that will be needed. If phones are the only devices that need to be supported, then transcoding middleware will be sufficient. This category of middleware can render the web pages in the appropriate format to match the devices' browser, such as transcoding HTML used on the web to wireless markup language (WML) or to handheld device markup language (HDML), which are used on these devices.

If intelligent clients need to be supported such that applications are written in a distributed design (distributed computing, client/server, peer-to-peer) then a message queuing or synchronization middleware will be needed.

What Type of Information do the Mobile Users Need?

Wireless users are generally mobile for some significant portion of their day and carry out their work while in the field. While they are mobile they have a variety of needs for business information. Sometimes they need to capture information, such as when an insurance claims processor visits the burnt-out home of an unfortunate policyholder. In such a situation, the insurance representative needs a variety of data: access to policyholder information such as policy information and payment history. The rep also needs to capture information (photographs of the residence and a list of damaged items), which becomes part of the claim.

Information requirements need to be thoroughly understood and not limited to historical capabilities. For example, a service technician may benefit from an installation tutorial when handling an unfamiliar refrigerator compressor. Their point of need is while they are at the job site, although historically, there would have been no technically viable means to deliver such a value added piece of information, today there is a suitable mechanism, such as a DVD player in a laptop computer. As networks improve in speed and throughput, a future way to solve this problem is to use the airwaves to deliver the required video.

In the requirements gathering phase, focus on what users could use to improve their performance. Identify inefficiencies in their work processes such as time wasted driving to a location only to find out that the repair truck was carrying the wrong equipment to repair the utility pole. Quantify how many of these workers are in the company as a means of building the business case for undertaking a wireless data project.

If the user's needs point in the direction of an intelligent device then either a synchronization or message queuing platform will be needed (or both depending on the timeliness of the information).

What is the Nature of the Interaction They will Have with the Data?

Mobile users have different levels of interaction with data while they are completing their business tasks. When they are reading data do they need to see it in a specialized format, such as a spreadsheet? How much and what type of data input is required? Is it data that can be captured through a structured means such as scanning a barcode? Or, is it the case that the data capture is akin to note taking, as would be the case of the insurance rep who is recording the lost items that are being told to him by the policyholder. The richness of data required to complete a mobile task will bias the choice of device. As the needs move to an intelligent device, transcoding middleware will not be applicable.

What is the Time Sensitivity of the Business Process and the Data Associated with it?

The time sensitivity of the captured data can also be used to define the type of middleware needed in an implementation. Some data is required in near real-time such as the package tracking information at FedEx. Their customers expect to be able to track the movement of a package from the moment it leaves the sender's location until it is delivered at the receiver's door. This level of information currency to run the business demands a message queuing wireless platform.

Where the data currency moves from minutes to hours or days, then a synchronization platform will become a likely candidate for wireless middleware.

Which Networks Need to be Supported?

Wireless networks have less than ubiquitous coverage. Some focus on cities, some on highways, and some, such as satellite, have large footprints but lack in-building penetration. This generally mandates that the application needs to be able to be delivered over more than one network so that users can be reached when needed. Middleware vendors need to be asked which networks they support. Some only support IP-based networks while the more experienced vendors support a very wide range of networks, even those that do not support IP.

How will these Factors Change in Version 2 through N of the Application?

Wireless applications are in their infancy and are guaranteed to morph as they mature. Consider the set of questions above and how the user and business needs will change over the next few years. The choice of middleware is a commitment to an architecture and a vendor. The trends in handheld devices indicate that increases in computing power are on a path that resembles the improvements charted for the PC over the past twenty years. This translates into having enormous computing power available to us while we are mobile. As will be discussed in the next chapter, wireless networks will improve at a slower pace than handheld devices. Network speeds will not be jumping by leaps and bounds the way processor speeds have consistently done. As a case in point, in the last twenty years processor clock speed has increased 1,000 fold while wireless network speeds have only doubled or quadrupled. This means that applications will continue to be designed to rely upon the robustness of the device to overcome the weakness of the network. This means that message queuing middleware will become the dominant type of wireless platform.

What to Look for in a Middleware Provider

The three types of middleware described earlier are supplied by many wireless middleware providers. If your needs are very specific, you can look at specialists, such as those that deliver web content to WAP phones only. However, to meet your long term needs, it is best to choose a middleware provider with a comprehensive platform supporting a broad range of devices, networks, and computing paradigms.

Prospective purchasers should ask the following questions:

- *What operating systems and transport protocols will the middleware product support?* Wireless device operating systems are growing more complex, and the number of options is growing. Be sure to choose a vendor that will meet your long term needs and shows a commitment to supporting emerging devices (e.g., J2ME, Microsoft's Stinger, RIM OS, Palm OS, EPOC, others).

- *Does the middleware support interfaces to enterprise applications and software standards such as SOAP, Java, and others?*

- *What special performance features does the middleware support – for example, least-cost routing, message prioritizing, error correction, and guaranteed message delivery?*

- *Does the middleware product support intelligent clients and web browsing?* Access to web-based applications is appropriate for some users and some applications but many applications are more effective when they utilize an intelligent client application that leverages the power of the device. Intelligent client applications provide a local data store and allow for offline operation when disconnected. This is becoming an increasingly important feature, if not an essential vehicle, for delivering mobile applications to end-users.

- *What security and encryption measures are built into the product?* Standard security and encryption algorithms are becoming increasingly available (and integrated) within middleware product lines.

- *Does the middleware enable synchronization with well-known server databases?* Some middleware products enable users to connect to the "big databases" such as Oracle, Sybase, Lotus Notes, DB2, SQL Server, and others. Many products provide a combination of middleware tools and applications development, which allow programmers to write apps and use a GUI builder and VisualBasic code.

- *Does the middleware solution support a variety of devices and important desktop applications?* Lotus Development Corp., for example, issued its Domino product as a server component of Lotus Notes that acts as middleware, wirelessly enabling mobile handheld applications such as e-mail, messaging, and calendars. The Domino solution offers these myriad devices a framework for messaging, e-mail, calendar, corporate directory, and address book.

- *Does the middleware solution support host applications?* Some do. For example, some products connect mission-critical legacy applications residing on IBM mainframes, AS/400 or UNIX systems to an RF network. This enables a mobile user simultaneously to access multiple legacy systems from the RF device using a technique known as server-side emulation, wherein the application resides on the server communicating with thin-client software on the wireless device. For example, a mobile worker in a warehouse scans a UPC code. Although the code means nothing to existing systems, the middleware product (in this case, Point Information Networks' WaveLink ActiveBridge) has an application that checks a cross-reference database to find an internal parts number. It looks up that part number on a host locator system such as an AS/400, then updates the inventory program on a UNIX system and sends the data back to the RF device in real time. The factory worker then slaps on the crate a barcode label printed from a wearable device on his belt. The crate is then delivered to the location indicated on the screen of his RF device. All of this is done simply by scanning the UPC code (Arielle Emmett, "The Focus is Middleware," in *Wireless Integration*, PennWell Publishing, September-October, page 22.)

Enterprise Examples: Middleware Optimizes the Wireless Connection

To give you some idea of the breadth and depth of corporate solutions using a middleware solution, consider the following:

- Reuters America, developers of Reuters Market feed, used a middleware solution to develop the company's first mobile wireless solution for 24 X 7 investors. The MarketClip service initially allowed mobile subscribers to access the Reuters network for news and stock quotes (but not actual trading; that was added later) on average 80 times a day per subscriber. Mediating between the client legacy network and wireless subscribers, an Aether Technologies middleware solution delivered real-time data feeds from Reuters America to subscribers. Aether-hosted servers in Maryland provide filtering and optimization, then delivery of the relevant data to wireless subscribers' Palm PDAs on CDPD or Mobitex networks. Application software on the client end talks through the middleware, which then handles the RF protocol conversions. In effect, the system is a corporate intranet extended out to customers by utilizing three public wireless networks.

- American Freightways (AF) based in Harrison, Arkansas, utilizes a Broadbeam middleware solution to provide compatibility between its terrestrial and satellite networks used to track for hire-motor carriers throughout 28 states. AF has outfitted 1,000 of its trucks with a complete wireless data system to meet demands for real-time communications between the drivers and the company's 221 customer service centers. The system employed the Broadbeam middleware to provide connectivity to both terrestrial and sky-based networks providing tracking coverage.

- MCI Communications (now Worldcom) began using Broadbeam's middleware (specifically ExpressQ) to support a mission-critical mobile dispatching application. The company had been looking for a wireless data solution for dispatchers and field service technicians. Before implementation technicians had to drive to an office to pick up printed work orders, then keep written notes throughout the day. Technicians returned to their office centers at the end of the day to enter the information into MCI's computer systems. If schedule changes were required, dispatchers would page technicians who responded by telephone. The system was deemed too inefficient and MCI custom developed an application Dispatch Manager to run in connection with wireless middleware. The Broadbeam ExpressQ product was used to support communications across a broad range of networks and operating systems, providing store-and-forward message function, logical name-based addressing, and automatic network switching. MCI estimated that it saved nearly $7 million dispatch and service through its wireless data solution. By eliminating two hours of administrative time for field technicians, MCI gained 25% more time to serve customers. The improved dispatch efficiencies of the system enabled MCI to reassign about 100 dispatchers and consolidate field force dispatch operations formerly performed at 80 network information centers into 11 regional "ONE centers." In the next chapter we will explore the numerous varieties of wireless data networks and mobile devices to guide you through the alphabet soup to better understand what meets your needs.

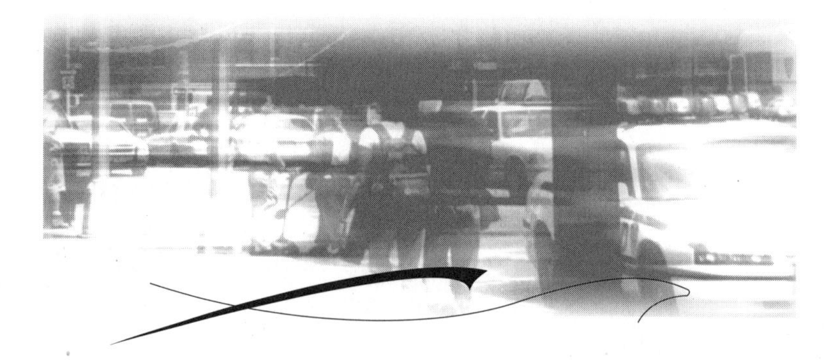

7 WIRELESS NETWORKS AND DEVICES COMPLETE THE MOBILE LANDSCAPE

ALL WIRELESS DATA COMMUNICATIONS TODAY are a trade-off between narrow bandwidth and large expectations. Managers must make choices between application richness and data speeds, latency (delays) and coverage, form factor (the quality and type of device) and ease-of-use. They decide on the use of proprietary, data-only networks (e.g., Motient's ARDIS network and Cingular Interactive's Mobitex network) or opt for the use of 2.5 Generation (G) and 3G networks supporting both voice and data such as those being deployed by carriers in Canada, Europe, Korea, and Japan. They consider satellite networks, which offer close to 100% geographic coverage, or footprint, while weighing the higher cost as compared to terrestrial networks. Consideration is also given to ruggedized devices vs. those commercially available and can the sales force truly use a wireless phone for all of their needs while in the field. Managers will also deal with the question of the wireless platform, or middleware, to devise an optimized solution for mobile users. To whom do managers turn for support and advice? How do they shield application developers from the variations and weaknesses of wireless networks, ensuring the right data gets through reliably on a variety of devices and that when network technology changes, their applications can readily take advantage of these changes?

Network Trade-Offs

Networks are sometimes baffling items, since for most people, a network is understood only as a tower along the highway or the band markings on a cell phone. Networks come in all shapes and sizes, with different design points and performance characteristics. Some are designed for voice and data (an inherent trade-off) while others are built for data only. A wide variety of questions stare network managers directly between the eyes everyday, including those that relate to cost, availability, coverage, data throughput, network management, the skills of their staff, security, and the ability to inter-operate with existing IT investments of the company. To get a bit deeper behind these questions, we will expand on each of these items.

- *Coverage:* The cost to a mobile carrier for building out a network is directly related to the number of cells that they need to support, therefore, operators look to maximize the utilization rates on each cell. This pressure on the operators results in operators selectively choosing which geographical territories to cover. It's not surprising that every mobile operator concentrates in high-density cities, but beyond the cities, operators have each tried to identify the next best pieces of real estate to cover. Each operator's coverage map is unique, and, because of the size of any country, especially the United States, it takes several terrestrial network operators to cover the entire country. Networks can be integrated now to provide something approaching nationwide coverage; satellite networks, circuit-switched cellular, and dial-up networks can be used to supplement terrestrial coverage in remote areas or for specific applications, such as mobile asset tracking.

- *Bandwidth and data speeds:* We are a speed hungry society. The processors in our PCs have jumped from being measured in millions of instructions per second to billions, and the pace of improvement is not slowing. We are impatient if our LAN-connected PCs don't return a web page in a split-second, and we expect that our e-mail is delivered instantly anywhere in the world. The comeuppance occurs when we move to wireless data where network speeds are measured in thousands of bits per second. Someone quipped that their network operator told them that they have three speeds on their network, "slow, slower, and stopped."

  Although not comparable to the blinding speed of the wired world, wireless networks are quite capable of delivering information required by today's wireless applications. Analogous to highways, as soon as the network throughput capacity increases, the amount of traffic on it will likely fill this capacity overnight. For the next several years, network throughput will be a bottleneck for wireless data and will cause a greater reliance on intelligent clients that can provide a distributed application, which reduces the amount of traffic that is passed on the network.

- *Application – big or small?* Enterprises should drive applications, not carriers. Carriers can get overwrought about inventing applications to suit the coming bandwidth glut; their motive is to drive usage by convincing mobile users they need color screens, rich graphics, high-speed connectivity while mobile, and other bells and whistles. Most mobile users today believe less is more – provided the applications are targeted, convenient, and work reliably.

- *Encryption and privacy:* Virtually all providers now must offer encryption to create virtual-provided networks for corporate accounts. Additional software may be required when carriers do not provide adequate privacy measures.

- *Costs:* Packetization vs. circuit-switching is a great cost divider. Other factors include bundling (with other telecom services, e.g., cable, DSL, fiber), big bucket purchases (purchasing large numbers of minutes at flat rates), and customer-premises infrastructure. Comparative pricing between public and private radio networks should be evaluated.

- *Performance:* Latencies, down time, and congestion are other factors to consider when choosing a network. Since mobile network operators can be restricted by local municipalities on tower locations and the number of towers, high density user populations can find that their mobile transmissions are delayed. Much like a highway when there are too many cars on it, congestion reduces the effective throughput of the mobile network.

Understanding Network Types

One of the fundamental differences between wireless data networks is how they are designed to move data. One architecture, *circuit-switched*, is built on the concept of establishing a unique connection between the two end points of the data call, just like a voice call is constructed on landline phones. The other approach creates multiple paths between the end points of the call and places "from" and "to" labels on each piece, or packet, of information that is sent. This approach is called *packet* networks. We will explore these two network types next. Refer to Table 7.1 for a sample list of network operators by network type.

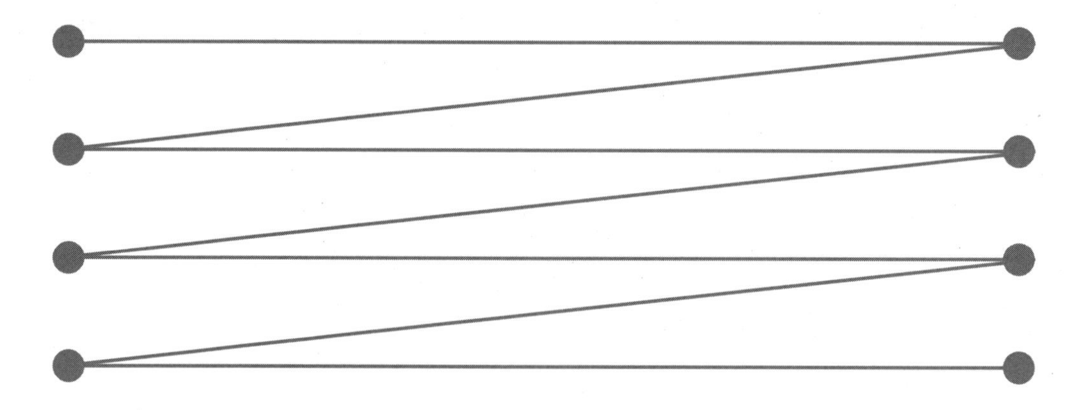

Table 7.1 Wireless Network Operators and Technologies by Network Type

Network Type	Network Operators (Network Technology)
Packet Data Networks	Cingular Interactive (Mobitex), Motient (ARDIS), various cellular carriers such as AT&T Wireless, Verizon and others (CDPD), British Telecom Wireless (GPRS), Nextel (iDEN Packet)
Circuit-Switched Networks	Sprint PCS (CDMA), VoiceStream (GSM), AT&T Wireless (TDMA), Verizon (CDMA), Nextel (iDen Circuit-Switched), British Telecom (GSM), Vodaphone (GSM), other Cellular/PCS carriers worldwide (various technologies)
Private Data Networks	Ericsson (EDACs), Motorola (Private DataTAC), Motorola (ASTRO), Nokia (TETRA)
Satellite	Wireless Matrix (MSAT Packet Data), NORCOM Networks, Qualcomm

*Note: this chart is not all inclusive, but provides a sampling of network operators by type.

Circuit-Switched Networks

Wireless wide area networks come in two different flavors: *circuit-switched and packet based.* Circuit-switched networks use a dial-up connection. For example, when dialing an Internet Service Provider either from home on a landline or when mobile on a wireless modem, users 'hop' onto a circuit-switched connection. Circuit-switched was originally designed for voice calls and allows network switches to connect a caller's request to a receiver's telephone. Since this action creates a roundtrip path for the conversation, or circuit, the technology became known as circuit-switched. This connection technology was carried into the wireless world of voice and then onto wireless data. Circuit data systems assign data transmissions to individual "channels" (e.g., specific radio frequencies) that are exclusive to the user. The user pays for the amount of time actually connected; moreover, the circuit must be established each time through a dialing sequence. A transmission tower, such as those that cover the landscape, have a physical limit to the number of circuits that they can handle at one time. This is why, sometimes when using a cell phone, you might hear a network message "all circuits are busy, please try again later" or find a message on your handset "network unavailable."

Circuit-switched connections, built to carry voice, consequently have three major disadvantages for data: capacity, cost, and connection. Because circuits must first be established and then "torn down" at the end of each transmission, network capacity is wasted in establishing and ending data transmissions. This overhead reduces the overall network capacity for handling data. A further reduction in network capacity happens as a natural consequence of using a circuit. Each data circuit has a theoretical capacity to transmit data, measured in familiar modem terms, such as 56 Kbps (56,000 bits per second). Unfortunately, this is an upward bound that is not achieved for many reasons. One primary reason is no data being sent. Why does this happen? Without getting into the details of how data is packetized, shipped, received, and acknowledged, it should be obvious that only one of these actions, namely shipped, refers to moving data from one point to another. Also, shipping and receiving data requires handshaking or acknowledgments, and this overhead subtracts from the data capacity of a circuit.

Costs associated with wireless data on circuit-switched networks can also be relatively high. First, as described, circuits are not the most efficient method for moving data from point A to point B, and this inefficiency relates directly to increasing the costs of moving data using circuit-switched technology. Secondly, circuit-switched network speeds generally operate at theoretical speeds of 19.2 Kbps, which is slow (1/3rd the speed of a standard modem operating at 56 Kbps). The actual throughput is much less than 19.2 Kbps when transmission errors are detected and numerous data resends are attempted. Also, consider the case where a data transmission is interrupted before its completion, as is the case when driving out of coverage or entering a building where coverage gets so weak as to lose the connection. At that moment, since the complete data set did not get acknowledged at the receiving end of the circuit, the entire transmission will have to be resent, meaning the user will have to pay essentially twice for transmitting the data only once.

Circuit data networks also have the problem of connection convenience. Since they have to have a "circuit" in place to communicate, they are not "always on," meaning the user has to take action when he wants to communicate, and others can't send information out to the user. This can take time away from the user's day if he has to constantly check to see if new information awaits him.

Packet Data Networks

It is not surprising that circuit-switched technology, originally designed to handle voice traffic, is not the most competitive technology for sending wireless data. To overcome the inherent weaknesses in circuits, a different approach was developed to move data, which does not rely on establishing a single path or circuit between points A and B. Packet networks, as their name suggests, are built around the concept of discrete chunks of data. Each chunk of data contains enough information so that the network can determine critical items, such as:

- the sender (done with a unique addressing scheme such as an IP address or a user name)
- the receiver (a different unique address)
- security (optional), and
- the packet number (eg., this packet is 11 of 73 packets that comprise the total message being sent)

By tagging each piece of information, an interesting result occurs – the path the data takes between points A and B is now no longer important or relevant. Some of the packets could move from point A to C to B while other packets could move from point A to D to B. At the receiving end, the packets are reassembled in their proper order (in this case 1 through 73). The reader may see the parallel to this description and how communication on the Internet works, as the Internet Protocol (IP) is also designed for a packet network.

Packet networks, since they don't require the user to create a circuit, have the appearance to the user of being *always on* and therefore, very convenient to use. When connected to a LAN, users are using a packet network and, as such, can send and receive data without having to establish a connection. Similarly, while mobile, a user who is communicating over a packet network does not have to take the time or effort to establish a connection as the device is always able to send or receive information while the device is on and the antennae is active. Research In Motion's BlackBerry devices for e-mail are good examples of devices that utilize packet networks for delivering a superb user experience. Users can receive or send e-mail while mobile. Many of these users have no idea how the wireless delivery system works since they do not have to interface with it. They spend their time working with the e-mail application, simply pressing Send or Open to access their mail.

Another interesting improvement intrinsic to packet networks is how bandwidth or capacity is used. The algorithms for routing data are quite complex and are designed to maximize the effective data throughput across the network. Since optimization is done at a level above that of a single user's needs, spare capacity can be allocated at any moment to improve performance.

Customer billing on a packet network is also done on a packet basis, generally measured in a terms that a user can grasp, such as megabytes/month. Under this pricing mechanism, users are only charged for the number of packets delivered.

Packet vs. Circuit

Although architecturally an improvement over circuit-switched technology, most packet-switched networks in the current generation are still too slow to deliver data at sufficient rates to meet the needs of network-hungry applications such as video files or complex web page downloads. The packet networks with the best geographic footprints around the world are Mobitex and DataTAC. Deployments of General Packet Radio System (GPRS) are beginning in Europe, Canada, and Asia, and these networks, often named 2.5G or "two and a half G" present the promise of speeds in the range similar to dialup performance or better. This means the wireless world is currently at the crossover point where connecting wirelessly will be as fast or faster than connecting on a landline, dialup connection. This is a truly historic moment. Although adults have lived with the reality of "wireless is slow," our children will have the exact opposite interpretation of the world.

In the U.S., There Are a Wide Variety of Public Data Networks for Mobility

The two most popular specialized data networks in the United States, Motient Corp.'s ARDIS (see Chapter 2) and Cingular Interactive's Mobitex network, have always utilized packet switching. Moreover, many cellular networks offer CDPD protocol, which provides a form of wireless IP as an overlay to analog voice networks. The principal sacrifice users of these mobility networks pay is data throughput since these networks operate at 8 to 19.2 Kbps. However, in the near future, wideband mobility networks will become more pervasive. As seen in Figure 7.1, we will see a generation of wideband CDMA, *CDMA2000*, GPRS, and other advanced networks, both circuit- and packet-switched.

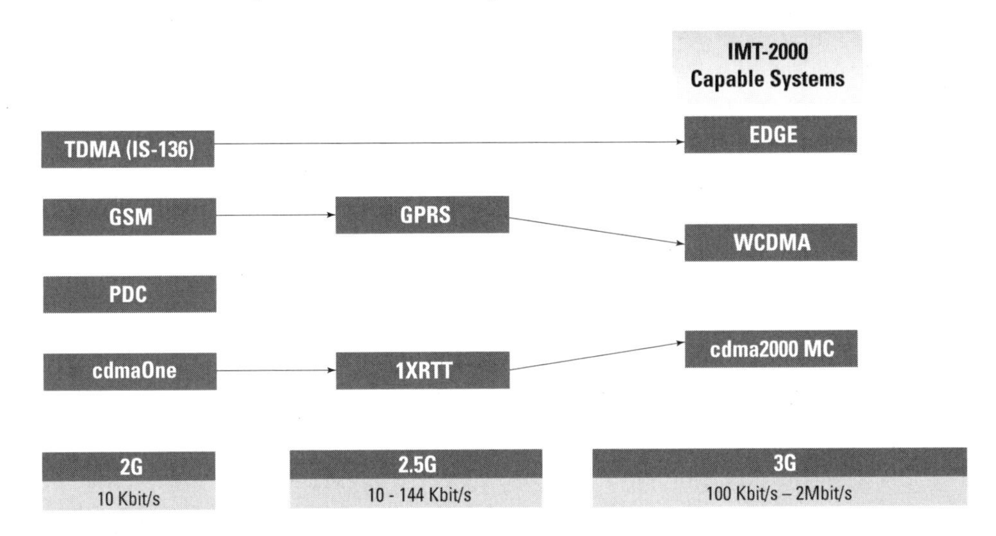

Figure 7.1 Wireless Data Network Migration Path

One of the most interesting publicly offered high-speed networks in 2001 is, rather was, Metricom Richochet, available in 13 U.S. markets as of spring 2001. This 128 Kbps wireless offering utilizes unlicensed spectrum and is faster than all other data-only public networks today. It provides ISDN-like speeds primarily for fixed wireless as a wire-line replacement for Internet access. The network's only real drawback is limited geographic scope (mostly to major metro areas) and price – $75 a month or more for subscription usage. Their other drawback was their inability to turn cash flow positive, and as a result, their growth has been stopped and they filed for Chapter 7 bankruptcy on July 2, 2001.

Thus far, the most productive and profitable wireless data networks have offered less speed and wider coverage for industry verticals (e.g., field service, warehousing, et cetera). Exclusive of the cellular and *Personal Communication Services* (PCS) services for which data is still a secondary function, the most important and widely deployed terrestrial wireless WANs in the U.S. are Motient's ARDIS and Cingular Interactive's Mobitex networks. ARDIS offers a maximum speed today of 19.2 Kbps, whereas the Cingular Interactive Mobitex network offers 8

Kbps with low latencies (transmission delays.) Another option, CDPD provides 19.2 Kbps and transparency to the ubiquitous TCP/IP protocols, meaning that applications written for delivery over TCP/IP do not have to be altered to run on a CDPD network. Although roaming is offered between CDPD networks, coverage is still limited principally to metropolitan areas. All of these specialty networks are best suited to short messaging, such as conveying financial transactions back to a central office, and remote monitoring, as they are not designed for large file transfer and multimedia applications.

What the Future Holds

Many analysts today believe that specialized data-only networks will ultimately give rise to packetized 2.5G and 3G networks – including GPRS, *EDGE*, CDMA2000, and various flavors of *wideband CDMA* (W-CDMA) – both being promulgated for advanced multimedia networks of the near future. The International 3G standard, known as IMT-2000, will incorporate and harmonize versions of both CDMA-2000 and wideband CDMA (a 5 GHZ spread spectrum technology that differs significantly from the North American offering). Carriers expect these networks will support a mass market for multimedia, including wireless voice, data, graphics, audio, streaming video, large file transfer, location services, telemetry, access to Internet and virtual private Intranets, and more. Essentially, with these networks, the anticipation is that users will gain the freedom of movement while having superb delivery speeds, making their mobile experiences enjoyable and practical for a wide array of applications.

In the meantime, the data world supports a wide variety of network types, including narrowband paging networks (the *Motorola Flex* and *ReFlex* networks are examples), and the *specialized two-way mobile radio networks* (SMRs), which support mobility. The mobility-oriented SMRs and the newly enhanced SMRs such as Nextel provide a combination of voice, data, and two-way radio capabilities; they offer cellular-like voice along with dispatch messaging and nearly nationwide coverage without roaming charges. Nextel's network enables corporations to operate a two-way radio system with cellular connectivity to loops of employees (something like a virtual private radio network). Other SMRs offer two-way dispatch services aimed at fleets of vehicles.

Private Packet Data Networks

In a private network, a company or government agency purchases radio frequencies and buys and operates the entire radio network infrastructure for the exclusive use of that entity. Enterprises may opt for a private network installation to ensure network availability at all times or to provide extremely high levels of security. Since the network is privately operated, bandwidth is not shared with other users as in a public network. Consequently, the network can be built to the capacity required to ensure availability and security.

A well-known private radio network is the one built by Federal Express (see case studies, Chapter 4). It's an excellent example of how wireless data can provide a competitive advantage by transforming a business model to one of high effectiveness and new levels of customer interaction and satisfaction. A number of private networks now exist for enterprises, public safety organizations such as 911 and police departments, utility companies, and government agencies. Many of these networks operate in the 2.4 GHz unlicensed band. Private microwave networks in the unlicensed band are commonplace – used, for example, to span large geographic distances while providing ample capacity for untethered enterprise/educational applications. The private network wireless technologies available today include *Ericsson's EDACS*, *Motorola's Private DataTAC*, and *Motorola's ASTRO*. *TETRA* is another emerging private network that is gaining market share in Europe and Asia.

Increasingly, though, private networks are being subsumed by higher bandwidth public networks with more efficient methods of boosting capacity (e.g., CDMA 1XRTT, GPRS, EDGE – see text below). Many enterprises are enhancing private networks with existing carrier offerings (e.g., CDPD networks) to boost geographic coverage and to decrease network costs.

The Rise of 2.5 and 3G Packet Networks

Digital networks are evolving rapidly – although not as rapidly as visionaries predicted even up to the first half of 2001. The push toward wideband mobility for wireless users has resulted in a competitive array of standards that require migration paths and harmonization within the core network infrastructure. The networks also require prodigious investment with uncertain returns. For example, five mobile telephony companies in the United Kingdom this year will be paying the British government $35 billion for access to Britain's third-generation radio spectrum located in the 2 GHz band. Germany followed suit with auctions garnering over $50 billion. In both cases, the mobile operators paid these sums to license the air; all of the costs of network build-out are completely additive to these costs! These bidding wars anticipate the wide-scale adoption of users who will gravitate towards mobile computing for every possible type of application for work and pleasure.

Internationally, the 3G standard is codified as *IMT-2000*. This standard will support wireless multimedia and large file transfers at theoretical speeds up to 2 Mbps per second. When deployed, IMT-2000 will harmonize two competing standards for 3G, the North American standard CDMA-2000 (an outgrowth of the Qualcomm-supported IS-95, used in current CDMA voice networks), and the European/Japanese wideband CDMA standard. Ultimately, consumers of wireless phone and data services should be able to "talk" and communicate data seamlessly across all 3G networks if harmonization is achieved.

Consider for a moment what 2 Mbps will mean. Although slower than most LAN connections, 2 Mbps will be able to deliver data at a rate that will make streaming video smooth. That means that teenagers who want to watch their favorite music videos will be able to request them from a web site and have them sent to them at a party or walking down the street. Clearly this is a far cry from where the state of the art is today, however, it is obvious why there is so much excitement surrounding the rollout of 3G networks.

2.5 G is the Reality Today

In the meantime, the first generation of 2.5G networks is beginning its commercial deployment. These networks, while not achieving the blistering speeds promised by 3G, will offer enough bandwidth for mobile users to support very rich business applications. The interim 2.5G standards include General Packet Radio Standard (GPRS), a 115 Kbps packet radio standard designed for GSM networks. Select deployments of GRPS are beginning in Europe and North America. Among CDMA carriers, the 2.5G standard known as CDMA 1XRTT will be deployed. 1XRTT will provide users with nominal data rates of 144 Kbps, enough to support graphics, audio, text, large files, and more. Another standard, *Enhanced Data For Global Evolution* (EDGE), considered an evolution of *Time Division Multiple Access* (TDMA), has promised speeds to 384 Kbps, but appears to be eclipsed by GPRS and wideband CDMA.

The EDGE standard was originally supported by AT&T Wireless, which now enjoys significant investment from NTT DOCOMO in Japan. In Japan, the most aggressive nation to adopt wireless data, wideband CDMA will be deployed over the next several years. In Europe, 15 countries in the European Union must begin IMT-2000 deployments, by law, no later than January 2002, and 3G licensing is already taking place in New Zealand, Germany, South Africa, and the Netherlands, according to officials at the Global Mobile Suppliers Association. Among most GSM operators – and there are over 400 in 42 countries (GSM now commands 450 million subscribers worldwide) – most are already planning on packet-based GPRS deployments for Internet support and high-speed wireless. As many as 80 to 100 licenses for wideband CDMA (IMT-2000) will be purchased worldwide. At the beginning of the migration, most GSM operators will go with GPRS, some will choose EDGE at 384 Kbps for data, and some will stop at EDGE and not move on to 3G. But others will wait for IMT-2000 because they're in the bigger markets and need the network capacity.

Satellite Networks & GPS

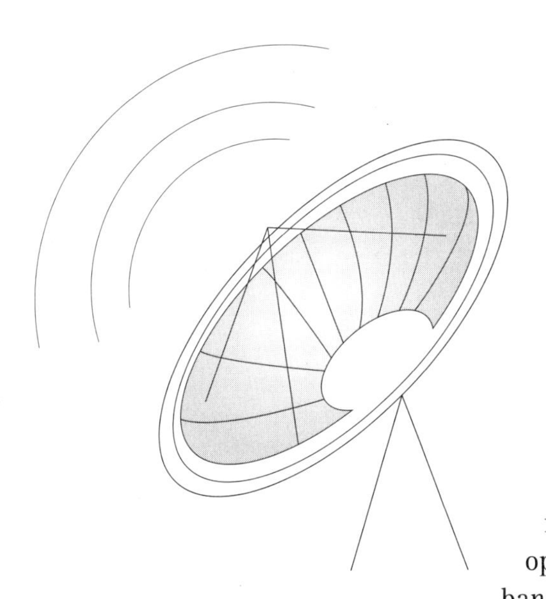

Satellite networks have always played a significant role in data transmissions worldwide. Satellite telephony and data coverage are most important in rural and geographically remote portions of the world – areas not otherwise reached via traditional land-based wireline or wireless networks. The cost of satellite networks has been high, and there is considerable diversity of coverage, speed, and pricing. Commercial satellite services can be divided according to the types of orbits they maintain and the frequencies in which they operate – for example, L, C, Ku, and Ka bands for commercial services (other satellite bands exist for government-only use). All commercial frequency allocations lie in the microwave region. For example, L band frequencies are in the 1-2 GHz range, and Ka band is assigned to the 38 GHz range.

As frequency goes up, satellites are more prone to atmospheric interference; however, modulation schemes and error correction can make transmissions at high frequencies viable.

Geosynchronous Satellites

Satellite services are determined in part by the orb's position with respect to the earth. *Geosynchronous satellites* (also known as GEOs) are the most common type, offer the broadest coverage, and have orbital speeds equivalent to the earth's own velocity of rotation. Orbits are as high as 24,000 miles above the earth. Three GEOs can supply global coverage, and network management is comparatively simple. Services from GEOs are reliable and offer network availabilities approaching those of fiber. But GEOs have higher latencies due to the length of signal paths, which presents a problem for highly interactive wireless data applications requiring fast response times and turnarounds. Still, companies such as Motient and Cingular Interactive have been successful offering a GEO satellite service to complement their terrestrial coverage.

Mid-Earth Orbit Systems

By contrast, *mid-earth orbit systems* (MEOs) position themselves a few thousand miles from earth, with orbital speed's faster than earth's rotation. They require more ground infrastructure than geosynchronous systems, are more expensive to manufacture, have greater mass, and must be radiation-hardened. MEOs have been most effective thus far in the government space, although some commercial satellite companies (e.g., TRW and ICO) have announced commercial deployments for this century.

Low Earth Orbit Satellites

The third type of satellite, *low-earth orbit* or LEOs, is the most controversial. They require very large constellations of satellites to ensure coverage, have limited footprints, and high rotational speeds. Located just a few hundred miles above the earth, LEO constellations have been used by Orbcomm, Iridium, and Globalstar. Orbcomm has enjoyed popularity for terrestrial data coverage, primarily to monitor and control fixed and mobile assets such as railroad cars. Iridium, which operated on a global satellite telephony model, suffered bankruptcy, and Globalstar is also deeply in debt (Alan Pearce, "Is Satellite Telephony Worth Saving," in *Wireless Integration*, September-October 1999, page 19). Price resistance to many LEO-based satellite services remains a serious problem; and as terrestrial networks become more robust, the problem could become more acute. Satellite vendors have struggled mightily to adapt and modify their business models to the price and applications squeeze, and some consortia (e.g., Teledesic, Hughes Network systems, and Loral are examples) are seeking markets in broadband satellite services, including the very high-end (terabytes and beyond) transmissions for scientific and industrial data. Beyond niche markets and government/scientific work, though, the future of satellite services remains uncertain.

In the enterprise/industry satellite space, narrowband offerings have included two-way paging and telemetry offered by LEOs, which use low-powered transmissions from small user terminals. Service providers such as ICO and Motient can offer bandwidth for both voice and low-speed data; however, they do not accommodate LAN connectivity or high-speed Internet access. By contrast, medium bandwidth systems offer more flexibility for applications. For example, *Very Small Aperture Terminals* (VSATs), which are named for the small dish transceivers located in ground units, have been popular to ensure simultaneous broadcast delivery of relatively large amounts of data. Utilizing C and Ku band spectrum, VSATs have enabled corporate LAN communications in Europe and North America, and basic telephone service in less developed parts of the world. Systems can operate in the multiple hundreds of kilobits per second on up to 1 Mbps range and higher – basically T1 speeds. Future broadband satellite services may provide high-speed Internet backbone. Again, the market for these services is not certain.

Global Positioning System

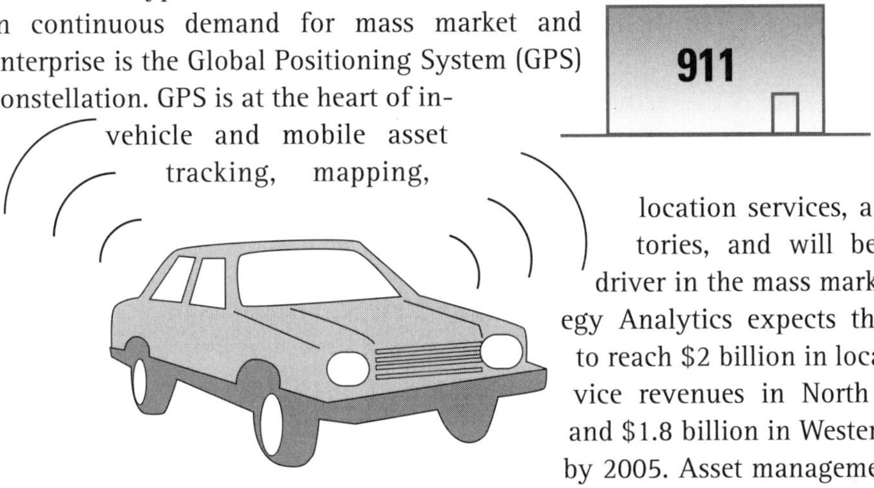

One type of satellite network that will be in continuous demand for mass market and enterprise is the Global Positioning System (GPS) constellation. GPS is at the heart of in-vehicle and mobile asset tracking, mapping, location services, and directories, and will be a chief driver in the mass market. Strategy Analytics expects the market to reach $2 billion in location service revenues in North America and $1.8 billion in Western Europe by 2005. Asset management applications utilizing GPS, a worldwide network of 24 satellites transmitting precise location data, have been available since the mid-1980s. GPS chipsets are being integrated into pagers, cellular and satellite phones, as well as trunked radios, and walkie-talkies. Some systems now unify the GPS chipset with a two-way radio using a single IP compliant processor, enabling a system to use the Internet and ubiquitous two-way paging networks to achieve real-time location data delivery (Brenda Lewis, "Location, Location, Location: The Killer App for Mobile E-Commerce?" in WirelessAgenda2000, Session #B6, Conference Proceedings, published by The Wireless Data Forum © WDF/CTIA 2000).

Wireless LANs and Fixed Broadband Systems

Wireless LAN architecture is focused deeply on the enterprise today. It is designed principally for networking and general business applications within a flexible enterprise space. Wireless LANs can be used principally to provide PCs with wireless connections to a corporation's computing infrastructure of servers, gateways, and web sites. The technology has become popular in warehousing and manufacturing, but could become an advanced form of white-collar office mobile office automation. Alternately, wireless LANs can be an integral part of the transaction processing systems with an enterprise. These LANs may interactively track the movement of inventory, assets, and value-added processing within the manufacturing facility, shipping dock, or warehouse, starting with data collection by the worker. In these applications, technicians use a handheld terminal – e.g., bar code scanners, digital image scanners, and/or a ruggedized handheld data collection device, which interfaces via a short-range piconet (Personal Area Network) to access points of the larger enterprise LAN. In this way, the wirelessly enabled LAN "brings" the interactive and transactional capabilities of the host computer to the worker's site. A combination of PAN and LAN capability helps direct and organize the transactions of the work process, aiding in workers' productivity.

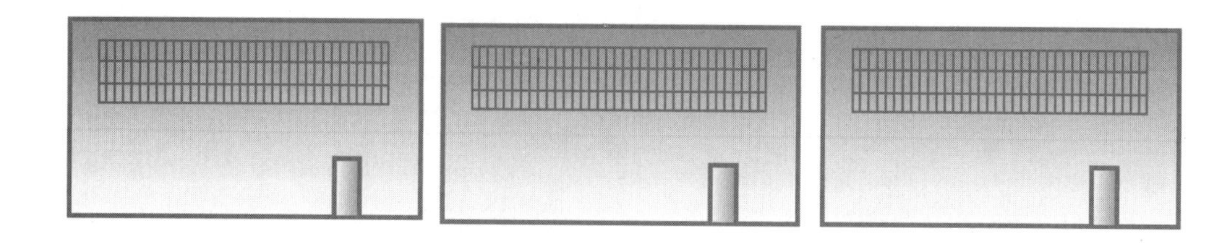

The deployment of LANs, principally within a building or campus environment, enables limited mobility and roaming. Wireless LANs use air links instead of cables to transmit data, most of which today conform to the *IEEE 802.11 standard*, pushing throughputs between 2 and 11 Mbps. Revisions of the standard have enabled throughputs as high as 50 Mbps or greater. Wireless LANs operate principally in unlicensed bands. They require a central controller, a series of base stations, and a network interface card (NIC) for individual terminals and computers constituting the "nodes" of the system. Wireless LANs feature either RF or optical connections between nodes and base stations, or from node-to-node or from base station to controller. Exclusive of a small subcategory of infrared LANs, RF-based wireless LANs permit limited portability within the workspace and allow automatic registration of nodes to the network as they are repositioned. This can be ideal in factory production areas, in universities and in health service environments where people and terminals move around constantly between buildings and rooms.

Wireless Bridges

A second category of LAN is the wireless bridge, which is used to connect a network segment or sub-networks. It is generally a single point-to-point link – a trunking or backbone connection at higher frequency, used to connect two physically separated nodes on the network (often in different buildings within a campus environment). Most unlicensed bridges operate at microwave frequencies, and at comparatively low power – from less than a watt to four watts at the 5.7 GHz band (other unlicensed systems operate at 900 MHz and 2.4 GHz). Licensed systems occupy the 23 GHz and 2.4 GHz bands, and some operate on separate bands for transmission and return path. Commercial wireless bridges (these are leased rather than owned) are becoming increasingly popular among enterprise customers. With the advent of licensed broadband wireless services, commercial wireless bridging over licensed bands in large metro areas will become more popular.

Broadband Wireless

In recent years the carrier market has introduced and reintroduced broadband wireless networks that require rooftop antennas and are useful as alternative forms of access. The three forms include *multichannel multipoint distribution service* (MMDS), *local multipoint distribution service* (LMDS), and unlicensed broadband wireless. The three types of networks can cover entire metropolitan areas. MMDS, also known as wireless cable, provides both subscription cable TV and high speed Internet access capability, although the majority of installations are one, not two-way. Commanding 200 MHz of spectrum, MMDS has more bandwidth to burn than PCS and cellular services; it can also operate with just one base station since channel reuse is not a necessity, thus minimizing infrastructure cost. But thus far the service has enjoyed only limited success. Its cousin, LMDS, has been designed for two-way industrial parks – Internet access and other high bandwidth applications. Operating at 25 GHz and up, commanding as much as 1 full gigahertz of spectrum, LMDS systems require a dense infrastructure of base stations and have been touted as high-speed Internet access alternatives to DSL and cable modems. LMDS installations are still spotty; as are networks operating in the unlicensed band (e.g., Metricom's Ricochet network). LMDS supports native Internet Protocol, as do unlicensed broadband carriers. None have found an expansive market as yet.

Wireless LANs, bridges and fixed, would require a book in itself – they vary by modulation system (e.g., spread spectrum or frequency hopping), transport protocols, network architecture (E.g., Ethernet or Token Ring), and price. A good source of information on Wireless LANs is PennWell Publishing's *Wireless Integration Buyers Guides,* 1999 and 2000 (the magazine is no longer published).

For the majority of this book, we will be focusing on wireless wide area networks using one of the networking technologies outlined earlier. Some consultants specialize in wireless LANs for campus environments only; others can integrate mobile and fixed wireless technologies together. Many enterprise IT specialists seek an integration between wide area and local area RF sub-networks. For example, police departments may use CDPD in-vehicle for data and voice transmissions and to connect with computers in central crime fighting laboratories and agencies; the units can be switched automatically to LAN access in-building (e.g., entering police headquarters) to accomplish high speed updates and file transfers.

War of Devices

Mobile devices are available in an expanding array of shapes, sizes, features, designs, and price points. It is rare that a product category is so varied and this speaks directly to the early stages of this market. It is simpler to deal with this panoply if we apply some sub-categories. The category can be subdivided into two large groups: those with standalone computing power, capable of running business logic, and those that require a server to execute business logic. Good examples of the first type of device are PDAs and PocketPCs since these devices are sufficiently "intelligent" to run a business application, have room to load the application, can access local data storage, and can capture user input. For want of a better term, this first category can be called "intelligent" devices.

The other category is known as "thin" devices since they lack the computing strength required to run business applications on their own as mentioned above. Wireless phones are the flagships of this category. Instead of having the computing resources themselves, which would add cost and weight, phones rely on utilizing the resources of computing power on the servers that are reachable through the wireless network. Each design, "intelligent" and "thin," have their usefulness and utility, however, the inherent architectural differences help define who will use them and for what purpose.

Thin devices simplify the deployment of a wireless application since the effort is concentrated in one spot – the network server. Once the server has been tested and switched live onto a wireless network, essentially all users can access this application immediately. For example, if a network provider installed a directory assistance application on a network server, all smart phone users could access this application, without having to provision any software on the phone itself. This is because the phone manufacturer loaded a browser on the device in the factory, and the user can point this browser to the 411 directory assistance application. This server-side application approach is quite powerful in reaching markets and users who do not want to take the time or do not have the skills to load an application.

Intelligent devices are quite the opposite case and require applications to be loaded on them. While this makes them quite flexible and powerful once the application is loaded, there is the initial effort required to achieve a successfully working application. Given the technical barriers to loading software and operational issues in making sure that the application is properly communicating with backend systems over the wireless network, intelligent devices pose a barrier to widespread consumer adoption for loading wireless applications. Returning to Figure 5.3, *Convenience and Desire = Adoption* users require convenience. In an early market, where skills are low, it is unlikely consumers will be able to or desire to load applications on their intelligent devices. This makes intelligent devices well suited for employee applications and a less likely target for consumer applications.

One Size Fits all vs. Multiple Devices

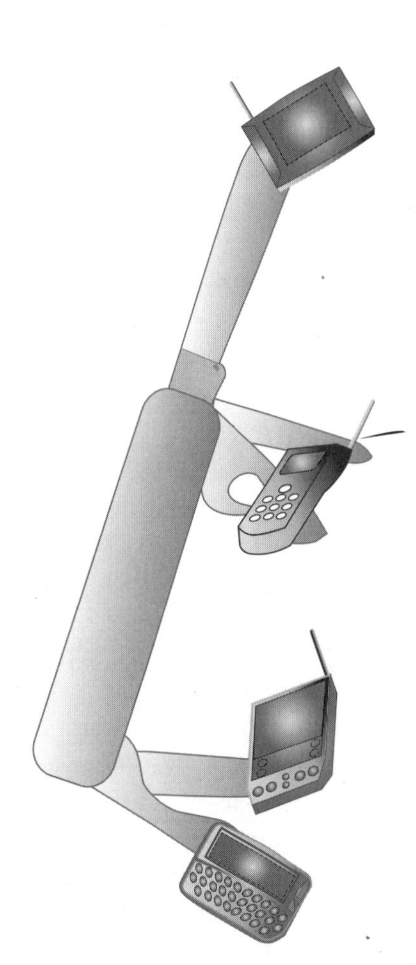

For the last several years wireless manufacturers and consultants have been framing the data discussion as a battle of devices and form factors. Some have argued the industry is heading toward a significant and highly capable single form factor that will satisfy the majority of user needs – the 'one size fits all' argument, incorporating the best of features and functions, including sophisticated displays, graphics, video, audio, data, rich applications, and telephony. Others have argued just the opposite, namely that creating an array of devices that support distinctly different aspects of communications – that is, separate phones, pagers, PDAs, and laptops, to name just a few – is the only way to satisfy discriminating users. There are wireless mavens who argue that a single device can't do anything really well; there are users who say many devices are cumbersome and expensive to carry and operate.

Both sides of the argument seem to have merit, but they both lack a quintessential factor of the equation – the human who will use this equipment. Strangely enough, articles and speeches describe device technology, but the discussions have curiously ignored the fact that devices will be accepted or rejected based on their ergonomics. The human-machine interface is a powerful boundary. Some machines have captured our imagination because they "work" so well. The original Palm Pilot or the RIM 950/957 pager with its easy to operate thumbwheel and menu system are excellent examples of winners. There is a raft of losers; designs so bad that their names drop from memory.

So the question of "one device" vs. "many devices" must be framed in the context of "what user." A taxi cab driver will likely use one data device. It will be mounted in the cab, and it will have the requisite functions that she needs, specifically a magnetic card reader for swiping credit cards, a GPS navigation system with a screen to get to any destination flawlessly, a printer for providing a receipt, and an input device of some type to describe the destination location. She will also have a voice phone.

This contrasts with the white-collar professional, who no longer wears a white collar, and will be the center of attention in the debate of "one" vs. "many." We assert that the following will happen:

Wireless Device Predictions

- Multiple device types will be created in the mobile device space. Users demanding device-specific functions as well as integrated devices will drive this. There is no technology space that we can think of that has converged. Technology diverges to meet the evolving and expanding needs of users.

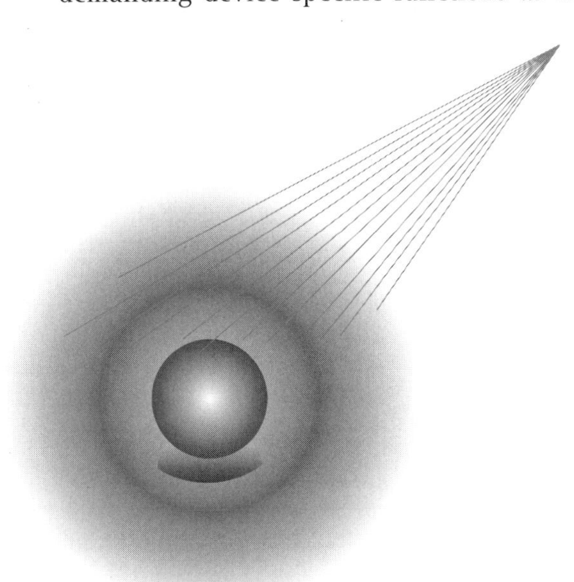

- Bluetooth has the potential to accelerate this expansion of mobile device designs since it portends complete convenience in interconnecting devices. Users could assemble a varying array of devices based on their moment of need and have them work together seamlessly. A PDA may be the device of choice for some parts of a worker's day, however, they may choose to switch displays to a heads-up holographic display to see more information, such as a spreadsheet. They may choose to do some typing and at that moment, prefer to unfold a keyboard for that purpose. Being able to pick and choose the range of function-specific equipment will make the user experience fun and productive far beyond the user experience any one device could deliver.

So, framing today's discussion of mobile and wireless devices as a battle between competing devices misses the point. It is not truly a battle between devices, but a battle of devices trying to capture the imagination of users. Devices will continue on their path of diversification and we will see many devices incorporate wireless connectivity creating tremendous variation and heterogeneity. Dick Tracy's wristwatch is not far from being a common item.

The Emerging Form Factor

What is likely to emerge? The major categories of mobile product today are the two-way pager, the PDA/palmtop and handheld PC, the laptop, and the smart phone. Each category has captivated a different type of user or a need of a user. For example, one of the most powerful and friendly wireless data appliances built purposefully for messaging is the RIM 950 interactive pager. This is a beautifully designed device with a tiny QWERTY keyboard for easy typing, a trackwheel that enables a user to scroll between functions horizontally, and a clear, iconic LCD display. The device enables a user to learn its main functions within just a few minutes – sending and receiving e-mails and pages, responding to and storing messages, accessing lists of message recipients, devising memos, filling a calendar with a daily or weekly schedule, and adding and editing a series of tasks. The pager is equipped with an intuitive address book function; moreover, it collects intelligence (making lists of e-mail recipients, contacts, and detailed addressing information) and makes repetitive tasks easier as time goes on. No wonder this device breathed new life into the executive e-mail/messaging marketplace. It is one of the best examples of a completely successful, ergonomically friendly product that does two essential things – mobile messaging and task organizing – very well. The RIM pager is the strongest argument yet that small devices with dedicated functionality may work better than amalgam products that try to be all things to all people. It is enhanced by the excellent packet functionality of the Cingular Interactive and

Motient networks; despite being narrowband, the response to send short e-mails is very fast and the user does not notice delays in either the sending or receiving direction.

On the opposite end of the spectrum is the "smart" device (that's a euphemism) that can't shoot straight. In smart phones today, the LCD screen is too tiny to display useful chunks of information; even "reading" news stories or stock reports off the Web is a ludicrous, sentence-by-sentence (even phrase by phrase) exercise. Paragraphs are split up. Readers are continuously interrupted by the circuit connection pausing to re-establish a session each time it must pull down new text. A smart button on the side of the phone is supposed to control most of the functions of the data interface, but users are frequently caught in "menu jail." Punching in letters to send an e-mail message is so awkward and time consuming (three short hits to get a "C" on the phone pad, for example) that users might as well just make a voice call. Strange function buttons at the bottom of the phone with abbreviations only Einstein might understand are the norm rather than the exception. In all, it is an example of "disruptive" technology in its earliest unformed stages – geeky, impractical, and ergonomically primitive.

Of course, not all "smart phones" in North America are so poorly designed, but users today are still trying to decide whether accessing the mobile Internet is really worth the fuss. Unlike the flexible c-HTML i-Mode phones, which offer Japanese subscribers wide accessibility to thousands of Web sites, most Web phones in America today provide limited Internet access and e-mail. Who wants to go to the trouble? Not the first wave of U.S. consumers, evidently. Buying books on Amazon with a cell phone – inputting credit card numbers while feeling intensely worried about security, lost digits, lost transactions, dropped calls – is enough to keep most consumers scurrying back to real bookstores. At the very least, most users will be tempted to run home and make that purchase on a fixed desktop device.

internet messages
disruptive technology
disruptive technology
lost digits lost transactions dropped calls
mobile messaging

PDAs to the Rescue?

Herein lies the true mobile enigma: When will the market develop a truly friendly data device, network, price, and content package, so easy to use and so compelling that users will want to bite?

Enter the revolutionary new class of devices – wirelessly enabled Palmtops, PDAs, and Pocket PCs. Most users are familiar with brand names at this point: Palm Computing's Palm VII, the Handspring Visor Platinum and VisorPhone, Compaq iPAQs, HP Jornada, Casio Cassiopeia, and more. These are basically data devices – pocket organizers/tablets with varying degrees of computing and calculating capability.

The classic "smart organizer," which first became popular with the Palm family of standalone devices, offered a proprietary operating system, calendars, addressing, list making, and now, wireless access to the Internet along with features known as "hot synching" – the ability to transmit and synchronize data to and from desktop systems. Some wirelessly enabled organizers (e.g., the Visor family) feature add-on modules and expansion slots that enable you to attach a cellular telephone, digital camera, or MP3 music player. The monochrome units are easy on batteries (many use replaceable AAA alkalines), and some (e.g., the Visor Deluxe) can connect with a Macintosh computer without an adapter to synchronize data.

By contrast, the Pocket PC or handheld PC category boasts a form factor slightly larger than Palmtops, but much smaller than laptops. These are truly handheld computers with many personalized functions. Many are equipped with tablets and a touch stylus, handwriting recognition programs that actually work (and can translate a handwritten message into a typewritten note), "virtual" (on screen) keyboards that enable typing, messaging and data input, well-organized menus, and bright color displays. These devices also utilize Microsoft's latest version of the Compact Edition (CE) operating environment; they can synchronize with Windows programs (not Macintosh) with Microsoft Outlook and Excel (but generally not with Microsoft Works, Outlook Express, or software from other companies). Pocket PCs generally come with a Web browser, voice recorder, and multimedia player, and the user interface is pure desktop. They are more expensive than Palm devices, but also more powerful and intuitive to operate and use (e.g., you can enter data with ordinary block letters, not Graffitti symbols, as is the case in the Palm systems). Aside from the Windows-based class of Pocket PCs, there are competing handhelds utilizing the EPOC operating system (developed by Symbian). These PCs offer a "checkbook" size computer equipped with a tiny, though usable keyboard. The OS includes a database and web browser and has the advantage of being able to accept and synchronize with a wide variety of software applications, not just Microsoft's. Virtually all Pocket PCs consume batteries at a rapid clip – for example, a Compaq lasts about four hours and the Casio models can last up to eight hours, which means recharging daily or often during a week.

"Ruggedized" Terminals and Hybridization

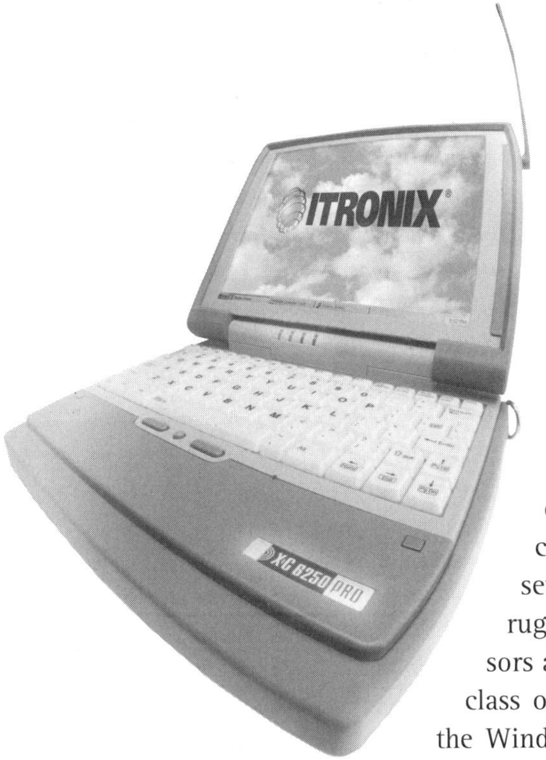

There is an entirely different class of handheld devices – ruggedized tools for specific vertical industries, combinations of bar code readers, and data collection terminals – that are acquiring a whole new range of capabilities. Some of these now accommodate voice processing (including voice over IP and voice to text conversion, so that workers can vocalize and "read" data into their computers), digital image and video capture, signature capture, transflective color displays, and radiolocation. A subset of these terminals – the "iron man," or ruggedized laptops, utilize Pentium processors and standard VGA screens. An emerging class of ruggedized devices also have adopted the Windows CE operating system, which eases programming for IT departments. Elements of ruggedization will become incorporated into the new generation of "hybrid" telephone-PDA devices. Most likely, all wireless devices will connect at least two ways – through the wide area cellular/packet networks; and through the wireless LAN networks (as exemplified by the 802.11 standard). As computing and communications come together, most handheld devices will be enabled through dual integrated modems both for wide area and local use.

Device Trends of the Future

Here are some likely trends in device form and features that are worth watching.

- *Evolution toward true multimedia messaging.* Messaging that involves the use of color, sound, text, and graphics is around the corner. When standards committees such as 3GPP were trying to define multimedia messaging standards in the past, most believed they would evolve from Short Message Services (SMS). Today, many see multimedia messaging as part of an evolution of the 'unified' messaging environment, in which users can send and receive all their messages on a common handheld device. The true multimedia system, rather than evolving from text-based SMS, will manage messages from voice mail, desktop computers, faxes, e-mail systems, pagers, in-vehicle systems, and more. The addition of video to this mixed media environment will be much further away, given the current course and speed of 3G deployments around the world.

- *More powerful device platforms – better browsers and application programming environments.* Devices will become increasingly more powerful and may incorporate multiple browsers, including WAP, HTML, and cHTML. Moreover, many PDAs will feature technology to support a JAVA-based programming environment (J2ME). J2ME creates a platform-independent programming environment enabling developers to create "clients" on mobile devices. These will drive wireless applications and servers to a much richer client/server environment. The client interface may center around a browser paradigm, however, the applications will either become distributed in nature (relying on the device for application logic and storage) or will use a browser in a store-and-forward approach. A store-and-forward browser allows users to access applications while off-line and then the device will push the application's completed forms to the server when network coverage is re-established.

- *Voice-driven Internet is likely.* Mobile users will "talk" their commands to a wireless phone or PDA or some combination thereof; new devices will be optimized not just for voice calling but for "calling" the Internet using voice XML. Voice recognition, text-to-speech, speech to text, will be an enormous improvement in the human-machine interface.

- *Paying for packets will become the standard billing model.* Wireless carriers will ultimately follow the Internet Yahoo! model for pricing; we'll see big bucket or flat rate pricing to encourage usage.

- *Embedded computing devices will be equipped almost ubiquitously with wireless capability.* We will become a telemetry-based society with data buzzing around us. Virtually all data collection, computation, and storage will be handled through an invisible, dense network of wired and wireless devices and host processors.

The ultimate "hybrid communicator" incorporating mobile voice, multimedia/Internet, and mobile data/transactional capability may only suffice for certain mass-market consumers. Given limitations in pricing, applications complexity, form factors, and networks, it is quite possible that multiple devices will be required by many – especially professional users – where optimization of certain functions are key. The future will bring us a wealth of creative devices. Take a twenty dollar bill out of your pocket, place your bet, put it in an envelope, and open it one year from now. It is going to be interesting.

project team
+ planning
+ research
+ requirements
+ vendor
+ testing
+ feedback
+ training
= SUCCESS
COMPOSITION
ABC
12

8 CREATING A WIRELESS BUSINESS: A PRIMER

PRACTICALLY EVERY COMPANY HAS BUSINESS processes that can be wirelessly enabled. Logistics, dispatch, customer service, sales, and marketing represent areas where wireless data access has already shown improvement in business efficiency and effectiveness. Many executives in enterprises are excited about the business gains their company could make if they changed a business process supported by wireless data technology.

Clearly there is desire – we hear interested executives representing practically every possible industry discuss this with us when we speak at industry events, technology shows, or business conventions. The most pressing questions they have are "should I start now or wait to use wireless data to improve my business?" and "if so, where in my business should I look for good returns?"

The first question has a simple answer, "start now." You will not be a pioneer since hundreds of companies around the world in industries from health care to insurance to manufacturing to public safety to fast moving consumer goods have all improved their businesses by using wireless data. The next chapter will deal specifically with where strategically to look in your company for opportunities. This chapter describes the mechanics of how to get started and represents more of the nuts and bolts of wireless data project planning and implementation.

The material outlined in this chapter will enable you to link with employees and subcontractors, coordinate mission-critical information distribution, develop direct connections with customers and suppliers, and boost customer awareness. Further, we will review the steps required to define your application and develop a set of goals and guidelines for becoming a wireless business. Finally, we present information on how to find appropriate wireless partners, how to build a Request For Proposal (RFP), how to develop prototypes, implement, test, and evaluate solutions. This chapter builds on knowledge we've presented in all the previous chapters and assumes some basic technical knowledge.

Getting Started

To get started you need to re-think your business operation with wireless – possibly to "disrupt" your current way of doing business and enter a new frontier.

Your goal is to reap significant return on investment (ROI) by:

- Increasing worker productivity
- Eliminating task duplication
- Improving customer service
- Providing point of service revenue opportunities
- Gaining a competitive advantage

Integrating wireless data represents a significant challenge for most businesses because they have not yet acquired the knowledge or expertise necessary to reengineer the organization. As we have discussed, developing wirelessly enabled business processes requires more than simply putting a desktop application on a mobile computer or PDA. Wireless is an entirely new paradigm of computing and communication. Here are the highlights:

Unlike a traditional LAN environment, a wireless application needs to contend with the harshest of environments – from varying coverage conditions to extreme weather conditions to narrower bandwidth to higher latencies. Wireless applications also need to address the unique needs of mobile workers. On the road, for example, mobile warriors need access to specific information – they find general surfing of the web unacceptable because the content offered to them is too rich causing downloads to take too long. Mobile users also want to use an assortment of devices and have information formatted appropriately for each. And, they need data to be synchronized between devices, so they don't have to spend a lot of time managing their data.

To avoid this dilemma, information technology and business managers can use the advice in this chapter to re-design the business process and architect the wireless application.

Steps to a Successful Solution

The following checklist represents an outline of the steps that need to be taken to ensure a successful rollout. Each step is critical to the overall success of the project and will be described in greater detail throughout this chapter.

- ☑ Assemble a full project team including IT, business managers, quality assurance, and most importantly, users.
- ☑ Develop a project plan allocating time and resources for each of the items below.
- ☑ Develop a specification of the business process by conducting on-site surveys and discussions with users. Be sure to include exactly what tasks need to be accomplished, what information needs to be communicated, what information users require, etc.
- ☑ Define the system architecture by developing requirements for the following components based on the business process specification. Submit Requests for Proposals to appropriate vendors, review responses, and choose the vendors that best meet your needs.
    - Mobile computing device/modem
    - Network(s)
    - Application
    - Middleware
- ☑ Develop migration, pilot, deployment, and solution support plans.

- ☑ Concurrently proceed with:
  - Procurement of devices and modems
  - Establishment of network connections
  - Establishment of proper test environment
  - Development of application
  - Staging of devices
  - Creation of documentation and training manuals
  - Training of support personnel
- ☑ Once developed, test the solution in the lab and, most importantly, in the field. Gather feedback from testing and make necessary changes.
- ☑ Train a small group of users and implement the pilot plan. Gather feedback and make necessary changes.
- ☑ Train all users and implement the deployment plan.
- ☑ Continue to gather feedback and improve/update the system as required.

Assembling the Project Team

Successfully deploying a mobile solution requires more than just developing an application. Successful project teams include representation from each business area involved in, or affected by, the development, deployment, and support of the solution. Table 8.1 below lists the different roles that should be included on the project team and a description of the role. Please note that a team member may take on multiple roles, but it is important to have each role represented in some capacity. Many of these roles may also be outsourced to companies with the expertise required, such as systems integration companies and hardware, software, middleware, or network providers that also offer professional services in these areas.

Table 8.1 Wireless Project Team Members

ROLE	DESCRIPTION
Project Manager	It is imperative that there is one person that maintains overall responsibility for the wireless project. This person will oversee, manage, and coordinate all aspects of the project from design through deployment. This person is charged with delivering a successfully working solution on time and on budget. This person possesses a blend of technical (telecommunications and IT) and business knowledge.
IT Manager	It is critical for the IT manager to be intimately involved in the project as the solution will ultimately need to be integrated into existing systems. In addition, supporting the system will most likely lie with the IT department.
Business Unit Manager	This person oversees departments receiving the solutions and must build the business case supporting the project. The business unit manager holds responsibility for the budget and objectives for the project and is critical in gaining and maintaining management buy-in for the project.
Operations Management	This COO-type individual will contribute to the project and make decisions regarding the overall impact on company operations. Thus it is vital that operations management is on board with the project, since this person can delay it if he feels it doesn't fit with the overall company objectives and operations plan.

▶

Table 8.1 continued

Role	Description
System Architect	The architect, who may also hold another role on the team, is responsible for designing the overall wireless application system. This role may be outsourced to an integrator, middleware provider, or VAR with experience in this area since the architecture must be correct for the project to succeed.
Systems Analyst	This person will review the system and look for ways to improve its design.
Application Developer(s)	When it comes to application development, organizations have a few choices: ▶▶ They can purchase an off-the-shelf product from one of many software vendors. ▶▶ They may hire a third party to build a complete solution for them or customize a partially packaged solution, or ▶▶ They can develop the application in-house.
Quality Assurance	Quality assurance testing is one common stumbling block to companies implementing a wireless solution. It is absolutely critical that the system be tested not only in the lab, but also in the field to ensure it is working properly before a major rollout. This involves a comprehensive QA plan that tests for the many different types of problems that may arise in a wireless solution.
Users	Users are a required part of the project team to provide input about their needs, describe operations in the field, test the system, provide feedback, and help train and introduce other users during rollout. When designing the system, it is also important that other team members follow users' activities for a few days to understand their business processes and with which information they need to communicate and have access.
Wireless Expert	Without having someone on the team who understands the ins and outs of wireless, the project is doomed to failure. This person provides knowledge of wireless networking and communication and helps design and develop the application and wireless communication components. Since this individual does not exist in most organizations, enterprises may outsource the function to an organization that has a solid reputation in the wireless industry.

▶

Table 8.1 continued

Role	Description
System Support/ Help Desk	A support team must be responsible for troubleshooting and correcting any problems that may arise with the system. This role is to field calls from users, assist them with difficulties, and escalate problems that cannot be easily resolved. The help desk team must be fully trained on the system and part of the initial design and development phases to gain a full understanding of how the system operates. They will provide input on features enabling easier diagnosis of problems and support for remote users. Keep in mind that in many instances, mobile workers may not be PC knowledgeable, and, oftentimes, the systems they are using are non-standard compared to the LAN environment. A crack help desk team is required to support a wireless application.
Procurement/ Purchasing	A project team member must be identified to be responsible for negotiating contracts with suppliers and purchasing system components.
Training	Effective training of users can make or break the system's success because if users don't understand the system, they will either use it incorrectly or not at all. The person responsible for training users on the operation of the system should be involved in the project from its inception to develop a full understanding of the system.
Technical Writer	This person is responsible for creating documentation for the mobile application and supporting software systems.
Legal Department	Someone from the organization's legal department must be involved to review and approve contracts and legal agreements with vendors.

Not every member of the team is involved in all stages of the project. The following chart (Figure 8.1) outlines when and how long each member should be involved.

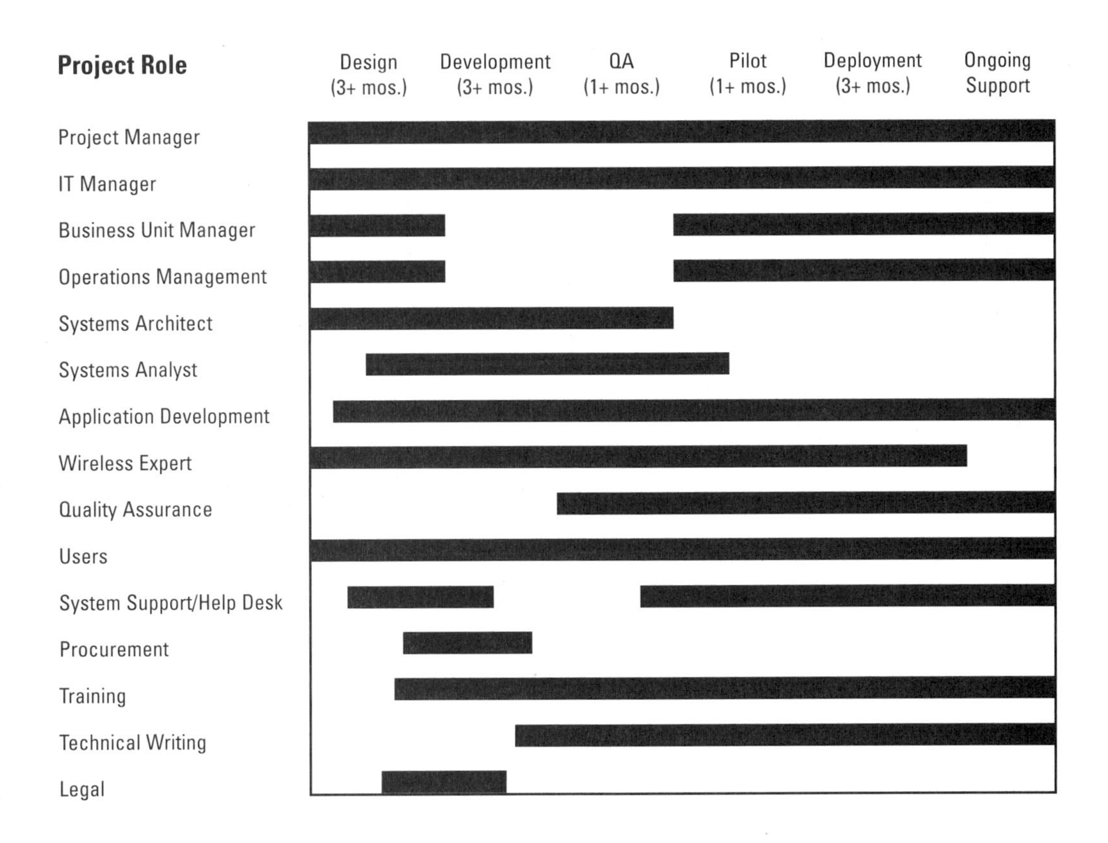

Figure 8.1 Wireless Solution Project Schedule and Role Assignments

Requirements Analysis: Developing a Process Specification

It is impossible to design a system that will improve productivity and enhance communication and customer service without first developing an in-depth understanding of the current business processes. Once again, this system is more than an application development project; it may, in effect, re-architect the operational processes within the business unit. This is just another reason why it is critical to make sure members from the operational staff, business unit management, and end users are all on your project team.

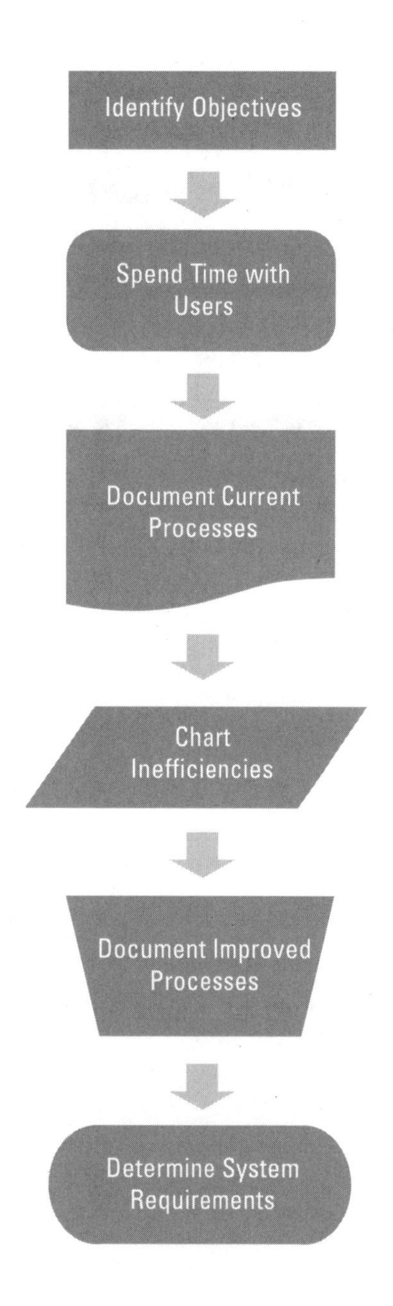

Figure 8.2 Process to Create Project Specification

The first step in developing a new business process specification is to identify operational goals for the new system and document the current business practices. Some of this information may already have been collected by the team driving the project when it developed a business case to receive management buy-in. These steps are sometimes blurred. For our purposes, we are assuming that management has already committed and bought into the vision and objectives of the project, and now it is up to you to implement it.

The process chart in Figure 8.2 depicts the various steps that will be covered in the next sections of this chapter.

Identify Objectives

Objectives for the project should be quantified, if possible, in terms of the following area(s):

- *Cost savings* – Examples: reducing the costs to service a customer; or to retain customers; or to acquire customers; or to distribute products.
- *Service enhancements* – Examples: tighter scheduling windows for performing services or having the right information available in real-time while meeting with the customer.
- *Revenue generation opportunities* – Examples: being able to reach new markets with new products or services; cross-selling existing customers with new products or services; or stealing market share by offering an innovative product or service.

Document Processes

The current business processes should be understood and examined in detail, not on paper but in real-life situations. The most effective way of documenting is by joining the mobile workers as they go about their day-to-day business.

During this documentation period, the system designers should determine:

- What information the user needs access to during the day including the complete inbound and outbound data flow;
- How the user currently gains access to information;
- Who needs to communicate and to whom, and how often;
- Which tasks are duplicated or done manually by in-house administrative workers on the behalf of mobile workers;
- What is the environment in which mobile workers operate (indoor/outdoor, rural/urban, etc.);
- Which resources will help the mobile worker better accomplish his/her job;
- Incremental revenue opportunities by performing services/sales at that point of customer contact;
- What information are customers/clients looking for;
- Problems encountered when communicating/working when mobile;
- What other devices, equipment, hard copy data, etc., the mobile worker now carries to perform his/her tasks.

This is just a starting list, which should be customized according to your specific business.

After completing this analysis of current business procedures, the operational team members should then document an improved business process designed to solve any problems identified and meet the overall objectives of the project. Senior management should review the business process – indeed, anyone involved in decision making, including business unit managers, key users, and other project team members to ensure complete buy-in. If everyone does not agree at this stage, there is a greater opportunity for the project to fail or not meet expectations at a later date.

Designing the System Architecture

The next step is to design the system architecture based on the business process specification and application requirements. This includes developing system requirements for each component of the system architecture including:

- Mobile computing device(s)
- Mobile/wireless network(s)
- Software application(s)
- Middleware
- Modem(s)
- Service and support requirements

Since many of these items are new to some wireless project managers, they are discussed in more detail in the following sections to help guide you through the decision making process. Middleware requirements were covered separately in Chapter 6.

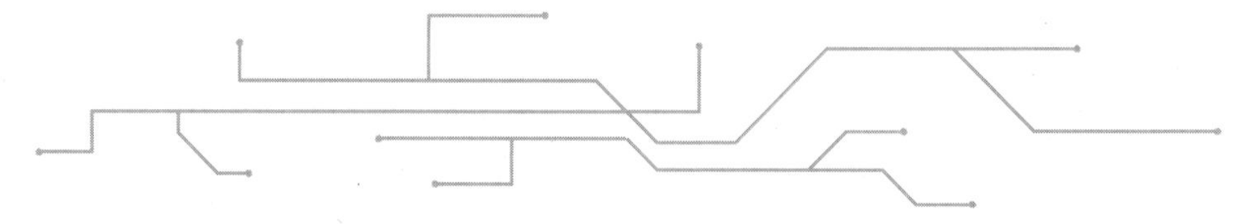

Issue Request for Proposals

After determining the system requirements, the next step is to identify potential vendors and issue a Request for Proposal (RFP) to qualified vendors. Good sources for identifying vendors include industry consultants, trade publications, peer referrals, and web portals. Be sure to clearly state your system needs and require responding companies to address how their product meets each need.

What to Look for in an RFP Response

When reviewing responses to RFPs, don't let cost be your sole determining factor. Oftentimes the most critical element is experience. That said, you also have to be aware of hidden costs such as training, consulting, or development fees. In addition to product specifications that are discussed in the following sections, there are many other factors to look for when reviewing an RFP.

- Has the vendor provided a system similar to yours before? Ask for references and call them! Look not only at the type of system, but also the size of the deployments. Does the vendor have any large-scale deployments? If not, be wary.

- What wireless experience does the company have? Have they been involved in projects from start to finish?

- What is the long-term viability of the company and the products? Is the company financially secure? Is wireless a key aspect of the company's vision and strategy? If it offers middleware or application software, you need to make sure it is committed to the project and you are assured that it will not abandon or discontinue support for its products in the future. If you are deploying a large number of users, you have every right to expect upper management involvement to show its commitment.

- What resources does the vendor offer such as consulting, application development, training, etc., and what are the related fees for such services?

- Are its offerings available today or are they a future promise? What is the upgrade path for its products? How often does it release new products?
- What are its customer support policies? What are its guaranteed response and problem resolution times?

Elements to Consider when Choosing a Mobile Device

There are many different types of mobile computers and devices, each with different features, functions, and purposes. The majority of devices are broken into approximately five to seven categories.

The categories and some major hardware vendors offering products are listed below.

- Laptop (Dell, IBM, Toshiba)
- Rugged laptop (Itronix, Panasonic)
- Clamshell Handheld (Hewlett Packard)
- Rugged handheld (Symbol, Intermec/Norand)
- Pocket PC or Personal Digital Assistant (Palm, Compaq, Casio, HP)
- Two-way Pagers (Research in Motion, Motorola)
- Smart Phones (Nokia, Ericsson, Motorola)

Unlike desktop computers, which are fairly homogenous, the purchase of mobile computers involves many decision factors, many of which are outlined and explained in greater detail in the sections below. Each of these criteria should be evaluated in terms of your company's specific needs and requirements to determine the best fit. Mobile device costs can vary greatly from a $200 phone or pager to a $7,500 rugged laptop computer. Therefore, you should carefully evaluate your needs, with an eye to scaling and expansion. Don't be quick to make a choice. Remember that the differences between thin and intelligent mobile device (discussed in Chapter 7) have a direct impact on the system architecture and application design. An error in device choice can be quite costly since it could cost system redesign and application recoding.

Durability

Durability is a critical factor, especially in relation to how the device will be used in the field. How well the device needs to respond to environmental elements such as stress, shock, temperature extremes, and water are dependent on the usage environment.

Some questions to consider are:

- *Will the device need to operate outside?* If so, you will most likely require a ruggedized device that is built to withstand all temperature extremes and is water-resistant. Also, be sure to test the quality of the display in bright light. Many screens are not designed for operation in direct sunlight and therefore may be difficult to read in that environment.

- *Will the device be vehicle-mounted?* If the device is vehicle-mounted, as often found in field service or public safety applications, the computer you choose needs to be able to withstand the stress and shock inherent in travelling along bumpy roads and high temperatures associated with unattended vehicles parked in the sun.

- *What is the possibility of dropping the device?* Must the device be 'hardened' or ruggedized for extreme weather conditions and wear and tear? The average mobile professional will drop his or her notebook computer from time to time; therefore, most laptops today are built to withstand a limited amount of stress and shock that occurs when dropped from a few inches onto a desk. But if you have workers that will have a tendency to toss their device into the back of a truck or bang it when wearing it on their belt, you will probably require a ruggedized device.

Form Factor

Yes, size does matter, but so does form factor. When you are mobile you want a device that is easy to carry and operate. Form factors are characterized by tradeoffs in size, weight, input mechanism, and ease of use. Each of these items should be prioritized to help determine the best overall choice. Which is most important, that the device be light, small, or larger, with easier data input? Remember, before a purchase, get the users' opinion too; they are the ones that will be operating the devices on a day-to-day basis.

Some items to consider when evaluating form factors are listed below.

- How much total weight can users carry? What other devices will they carry? Will they be carrying the wireless device all-day or only intermittently?
- Does the device need to fit in a briefcase, on a belt, in one hand, etc.?
- What input mechanism is best for your users? Non-PC users may find that a tablet device with an easy-to-use touchscreen will be most convenient. If your users wear gloves, be sure that the input device accommodates this. Other workers, such as delivery personnel, may be holding packages with both hands and have only a thumb to input information. If there is a lot of text to input, most users will prefer a full-size keyboard. If users are sending short messages or bits of information, the hunt-and-peck on a two-way pager or PDA may be suitable if size is the overriding design criteria.

Battery Operation

For many organizations, battery life is one of the most critical elements when choosing a device. Once again, it all depends on your users' needs.

Issues to consider when evaluating the battery operation include:

- How long will your users need to operate devices on their own power source? This will determine the necessary battery life of the device.
- What types of batteries does the device operate on? Many small devices such as pagers and PDAs operate on standard AA or AAA batteries that are inexpensive and easily stocked.
- What adapters are available? Can they plug into standard automobile cigarette lighters like most phones? Does the device offer in-cradle recharging? What is the recharge time required?
- If used with a wireless modem, does the modem utilize the device's battery or have its own power source? If it uses the device's power, add in extra time to the required battery life for normal operation.
- How large or heavy is the battery and charger?
- If the device was vehicle-mounted, can it utilize the vehicle's power or battery?

Thick vs. Thin?

A choice many wireless developers are facing today is whether to deploy an intelligent client that has a client application and local data store on the mobile device, or whether to implement a browser-based application. Both have advantages and many companies will find themselves needing both to support different user groups. If your users only need access to their data and application when connected, and downtime due to coverage holes or powered off devices is not an issue, then a browser-based solution offers many advantages. Web-based applications are easy to implement using rapid application development tools. Also, devices are inexpensive and easy to manage. In fact, you can probably leverage devices the user is already carrying such as smart phones or PDAs.

However, if the ability to operate offline and to push important information out to users (such as new work orders) is critical, then an intelligent client is the better choice. For the most effective intelligent client application, use middleware that supports store-and-forward message queuing and push message delivery.

Operating System

Even once you have made your choice between a browser or intelligent client, there are still more choices to make regarding what operating system works best for your mobile users. When choosing a device, you need to consider the power of the underlying operating system, as well as the compatibility with existing applications. If you plan on supporting multiple devices, your best bet is to use mobile middleware, which provides platform independence, allowing your application to be easily ported to varying devices.

Some issues to consider include:

- How robust is the application? Does it need a full-powered 32-bit operating system to function properly?
- What are the application's memory requirements? Many devices with proprietary operating systems offer only small memory capacity. If you are not willing to re-engineer your application for the mobile device, you will require a more robust operating system.
- Do you need to run multiple applications simultaneously? If so, the proprietary operating systems or those devices running DOS (which many still do because of the low memory requirements) are not for you.
- What kind of graphics do you require? Sometimes, the best mobile application is simply text based, since it delivers a rich enough user experience and is the easiest to design, code, and test. If a richer user experience is needed to display or capture the appropriate information, such as multi-tasking or windows, then you will have to consider an operating system that supports this level of function.

Application Requirements

Before making a final decision on a mobile computer, you will first need to understand what requirements your application(s) will have in terms of CPU speed, RAM, hard drive memory, operating system support, etc., to be sure that your desired choice meets those requirements.

Modem Support

Many mobile devices are designed specifically to work with certain wireless networks and have packaged modems to support those networks. Others offer the option of choosing from various packages depending upon your desired network(s) while some make it your responsibility to find modems that interface with your chosen network(s).

Following are some questions to consider when deciding which type of modem support you require in a device:

- *Do you want the modem to be internal or external to the device?* Wireless modems come in a variety of forms. Integrated modems are inside the device and therefore offer an integrated package that doesn't require any separate pieces or cables that could be easily lost or damaged. Conversely, external modems, which are offered in either the form of PCMCIA cards or packaged separately, allow you to switch modems if you are encountering problems, wish to upgrade, or decide to switch networks.

- *What level of coverage do you require?* Mobile modems, such as vehicle mounted modems that are "mobile" but not "portable," tend to offer better network reception due to their higher wattage output and externally mounted antennas, as compared to portable, internal, or PCMCIA modems.

- *Does the modem use the device battery or have its own?* In some instances, it may be preferable for the modem to have its own power source, which is easy to maintain with standard 9-volt batteries, rather than to draw from the device's battery supply.

- *What information regarding the modem is required by your users?* For example, do they need to be able to determine their battery level or network signal connection at a glance? If so, then you require a modem that offers these indicators via LED displays, or makes this information available to the application or middleware, which in turn will display this data on screen.

- *What cables or connectors are required?*

- *Will you be able to support multiple networks with the device?* For example, if you want to provide backup or nightly access via a wireline dial-up connection, you may want to consider using a device with a built-in modem and a PCMCIA slot for a wireless modem.

Peripherals

You should also consider what peripherals your users require or will make their jobs easier. For example, it will be much faster and easier to obtain accurate data by using bar coding and a barcode scanner within the mobile device than requiring the user to input a code. If your application is ideal for bar coding, you should consider a device with a built-in scanner.

Some applications also require connection to a printer. This could be a small, wearable printer or a device that resides in a vehicle. Connecting to printers and other peripherals will be much easier using Bluetooth technology (if and when the chipset becomes low cost), which allows devices in close proximity to communicate wirelessly, acting as a cable replacement. If peripherals are a requirement, you may want to consider choosing a hardware vendor that is planning near-term support for Bluetooth and understands the timing and cost of their offering.

Price and Support

Finally, it is critical that you consider price, warranty, and technical support during the decision-making process. When choosing a vendor, inquire about upgrade policies, future product pipeline, backward compatibility, support plans and possible leasing options.

Issues to Consider when Choosing a Network Provider

Just as there are many mobile devices to choose from, so are there network providers. The different types of networks were reviewed in Chapter 7. This section provides some decision factors to consider when choosing a network provider. When developing a request for proposal, be sure to include each of these criteria in order to make an informed decision.

Coverage

Coverage is an obvious issue but one that goes much deeper than simply looking at a coverage map. In addition to making sure the network providers offer coverage in all the areas that you require, you must also determine the level of coverage that they offer such as their levels of in-building penetration.

Some areas to consider are listed below.

- Do they have enough base stations in heavily populated areas to handle the volume of traffic? There could be a circle on a coverage map indicating coverage, but what is the level of that coverage? Test the signal strength in various areas in which you require coverage to determine that the coverage provided is truly adequate to communicate effectively.
- What type of in-building penetration do they provide (if you require in-building communication)?
- What is the effective throughput of data for message sizes that approximate your expected usage, e.g., if you plan on sending large messages (~ 500 Kbytes) do not accept performance data based on small messages (< 1 Kbyte).

Speed

The most important thing to realize when considering the speeds of wireless networks is that they will always seem too slow. Ask for performance benchmarks then use these benchmarks for designing your application. Consider your user population and answer the question, "what is a reasonable amount of time that they will wait for data to be delivered when they request it?" Use the benchmark data to determine the amount of data that you can send in this period of time and use this as the upward bound when designing the application.

Also realize that there is a lot of hype in the market in terms of network speed. Faster networks are coming, but wireless is working well in many applications around the world today. There is no need to wait for the promise of faster networks.

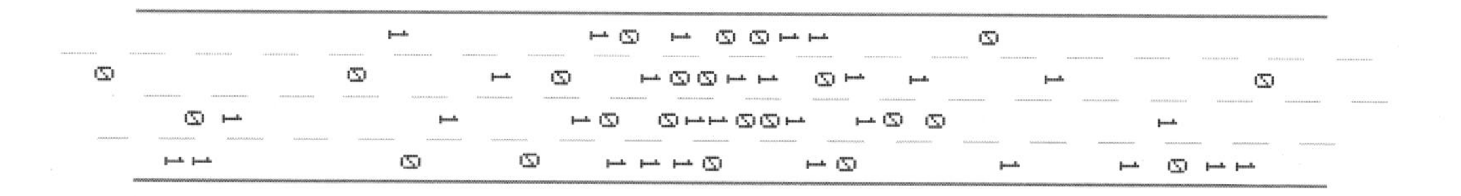

Network Capacity

One of the issues mentioned previously in relation to speed and coverage is network capacity. It is important to probe beneath the surface to determine the capacity of the network in your critical coverage areas and at what percentage of capacity the network is currently operating. Also ask what capacity is being used during high points of the day when your users will need to send their data. Network congestion has peaks and valleys, similar to highway congestion during rush hour. By understanding the conditions of the network congestion, you can make a more informed network choice and, possibly, design your application to pass large amounts of information during times when the network isn't very congested.

Reliability

Obviously a key issue is network reliability. If your mission-critical application is dependent upon getting data through, you need a very reliable system and network partner.

There are a few key issues to look at when evaluating reliability:

- What kind of Service Level Agreements (SLAs) do they offer? What is their guaranteed uptime?
- What is the percentage of network downtime?
- What does your partner's network operations center look like? What monitoring and troubleshooting tools does it use to support the network?
- Does it have redundant servers and connections to the main operations centers?
- How quickly can it fix things when something goes wrong?

Latency

Network latency refers to time delay between submitting a message and when it is received by the recipient of the message. Latencies can be affected by many different factors, including congestion, poor coverage conditions due to natural elements or obstructions, etc. These are impossible to calculate and will vary over time. There is, however, certain latency inherent to the network technology that can be evaluated on a consistent basis. The latency may not be a strong enough reason to dismiss a certain network, but it should be a factor to consider.

Host Network Connection Options

When evaluating network providers, you should determine which types of host network connections they provide. In other words, how do you connect your wireless server to the network? Via the Internet? Through an X.25 connection? What options are available?

Look at each option in terms of the following issues:

- How long does it take to set up?

- What is the cost to establish and maintain the connection?
- How much traffic does the connection support?
- Is there a limit or recommendation of number of users supported by the connection?
- What is the reliability?
- What is the backup plan in case the connection goes down?
- What information or monitoring can be done of the connection?

Complementary Networks

When choosing a network provider, you should also consider using a complementary network to supplement coverage in low or no-coverage areas, providing a back up in case the primary network goes down. Some networks offer complementary network services, such as satellite, within their product portfolio. Depending upon what you require, the most common complementary networks are satellite, circuit-switched cellular, and wireline dial-up.

Technical Support

Beyond the technology issues, there are always the business issues to consider to ensure that you get the best product, with the most support, for the best price. For such a critical link in your wireless data solution, your network provider's technical support needs to be world class.

Be sure to ask:

- What are its technical support hours?
- What is its response time?

- What is the escalation procedure?
- How are you notified if the network is experiencing difficulties?
- What is the network disaster recovery plan?

Cost/Pricing Model

Cost is an important consideration when choosing a network carrier. Pricing models usually fall into one of three categories, either buckets of minutes in circuit-switched, by consumption or flat rates. While flat rate pricing plans may make budgeting easier, you may end up paying more than you need to. The most important factor in considering different pricing models is to truly *understand how much data you will be sending.* To do this you must have a measure of how much data, in bytes, will be communicated daily. When comparing pricing, also look at the overage charges, roaming charges if applicable, and any other connection fees that may apply.

Reporting

Investigate what types of reporting and statistics the network operators can provide. Sometimes this information can provide great insight into how much traffic is generated, peak times of the day, average message sizes, etc. Plus it helps you analyze your airtime expenses!

Things to Consider When Developing a Wireless Application

Mobile application development is a different paradigm from standard application development because the mobile application will communicate with the back office in a disconnected fashion. Mobile users will also communicate using a variety of devices, sometime even multiple devices. Therefore, information needs to be formatted and filtered appropriately for each device.

These are issues that developers need to consider and accommodate for when developing a mobile application. Some organizations may choose to purchase a packaged application from a third-party software vendor instead of building it themselves. Mobile middleware can also be employed to handle many of the information retrieval, filtering, formatting, synchronization, and optimization issues.

Developing an Applications Checklist

The first step to application design and development is creating an application checklist to ensure that all major issues are considered.

Here are some key questions to ask while designing your application.

- ☑ How much data will you need to communicate?
- ☑ Does the application need to be always connected?
- ☑ What information does the user need?
- ☑ Can users benefit from access to additional applications such as e-mail or intranet access?
- ☑ Is the application highly interactive and "chatty," (two-way), or primarily one way? Some applications relying heavily on two-way communication are comparatively slow over wireless networks because of higher degrees of latency (delays).
- ☑ Is the application suitable for IP? If so, choose a wireless network that supports IP. For example, CDPD and GPRS networks are IP-based.

- ☑ What kinds of latencies, speeds, and performance parameters must the wireless application satisfy? These factors must be investigated closely before you decide whether current wireless offerings are fast enough and robust enough for your mobile requirements.
- ☑ What interfaces and mobile platforms are required? Today, wireless data often requires a microbrowser solution (designed for 'smart' devices, such as smart phones and PDAs), so that content must be written either in Wireless Markup Language (WML), Handheld Device Markup Language (HDML), or XML. However, the new generation of more capable wireless devices may provide new options for content developers; some content may be developed using standard Web-based HTML and voice XML (which uses voice commands to drive the interaction with the wireless platform).
- ☑ Can middleware improve performance or provide required features?
- ☑ What devices will the wireless application run on?
- ☑ How will billing and carrier services affect the decision to implement the application? What is the optimal structure for billing?

Information Delivery

Most mobile devices, with the exception of full laptops, offer small screens to display information. Still others have limited input mechanisms such as tiny keyboards, graffiti handwriting, or numeric keyboards. Information should be delivered to these devices in such a way that it is very easy to read and respond. For example, your application should include many built-in standard responses that are a one-key operation such as "Job Complete" in a field service application. Many devices offer software development kits to help you develop the user interface of your mobile application.

Understand that users will want to receive and interact with information differently depending upon the device. For example, some people may use smart phones for voice and short messages, Palm organizers for schedule and contact information, and laptops for e-mail and other office applications. In this instance, they don't want to read full e-mail on their smart phones. Your application needs to be able to filter information appropriately for each device. One way to do this is to send only short messages and e-mail headers and address information to devices such as a smart phone, interactive pager, or PDA. Depending upon the importance, the user can then decide whether to request the full message on that device, wait until later to receive it, or use his laptop to retrieve the message. You may also wish to include the capability to prioritize messages so that the most important message gets to the user first.

Multiple Devices

As mentioned previously, many users will wish to carry multiple devices for different purposes, or you may choose to deploy different devices to workers with different roles within your organization. How will your application handle communicating to multiple devices within the field force or per person? And how will it handle a user connecting with one device one minute and another the next?

Your application needs to include *some intelligent routing capabilities* to be able to recognize what device and network the user is on and switch easily between devices and networks. By developing *user profiles*, you can determine which devices users are carrying and what information they wish to receive on each device. When users connect or log in, you can then determine what they are currently using and communicate appropriately.

Data should also be synchronized appropriately between devices. For example, if a user receives e-mail headers on a two-way pager and chooses not to view the entire message, the message should be saved so that he or she can then access it later using a more appropriate device such as a laptop computer. An important issue for enterprise users, particularly mobile professionals, is the elimination of multiple addresses. Mobile users want to access the same information from each device. They want information to be synchronized between devices. They do not want to maintain multiple addresses for different devices. This is another area where the intelligent messaging capabilities of the back-end server come into play.

Push Technology

A critical component of a successful mobile application is the ability to push data to users. Remember that mobile workers are operating in a disconnected fashion and often roaming in and out of coverage. Therefore, they cannot be expected to check constantly to see if they have any data or messages waiting for them. *Information should be pushed to mobile users.*

Push can be done in a variety of ways. For example, the message could be pushed to all devices carried by the user since the host application won't necessarily know which one they are using. Or, if the user is connected, information could be pushed to that device. Finally, information can be filtered and pushed to different devices as discussed in the prior section. However it is accomplished, the ability to push is a key function of mobile applications.

Coverage Fluctuations

Network coverage is continuing to expand, but users will always experience coverage holes or fluctuations in coverage levels. Just compare it to your experience using a mobile phone while driving – conditions often vary. A variety of factors can influence coverage including rain, tall buildings, and hills and going deep inside a building. Virtually all mobile users will experience these conditions, and your application needs to be able to handle users moving in and out of coverage.

One way to accommodate changing coverage is to implement *store-and-forward message queuing*. In this instance, messages are stored in a queue when the user is out of coverage or offline and then pushed to the other side once the connection is re-established. This ensures that messages are guaranteed to be delivered as soon as the user is back online. As you can see, a good implementation of store-and-forward goes hand in hand with push technology. Mobile middleware technologies, such as Broadbeam's ExpressQ™, a component of its Axio™ wireless platform, provide these functions for mobile application developers.

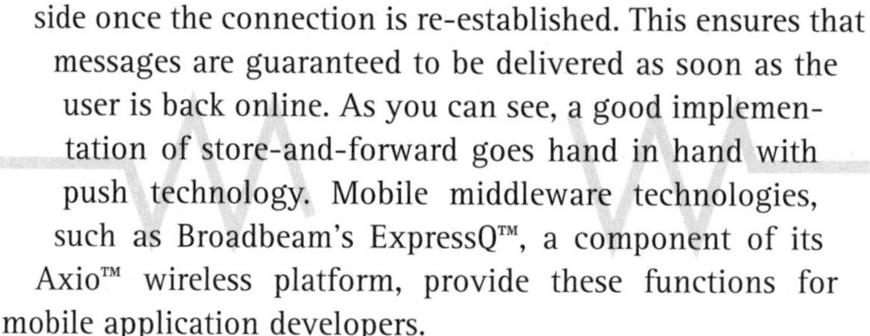

You may also want to consider adding features that allow the application to automatically reconnect to the network if the connection is dropped and resume transmission where it dropped off. You can do this by maintaining a virtual connection with the application so that once the actual connection is automatically re-established, the application can resume without having to restart or resend. This too can be accomplished using mobile middleware products or transport protocols that have been optimized for wireless.

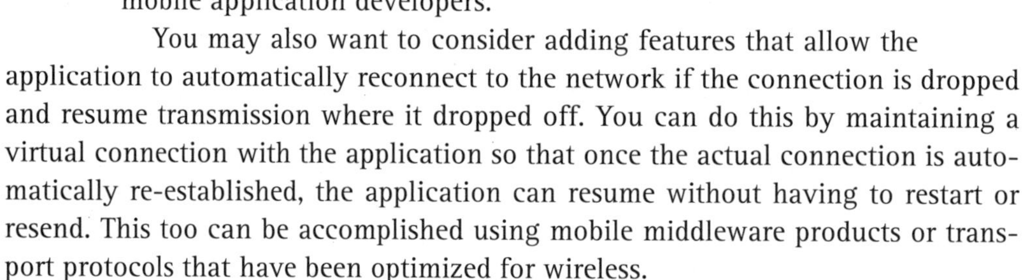

Bandwidth

As you know, wireless networks currently offer much narrower bandwidth than traditional wired or even dial-up networks. This will change in a few years with the delivery of next-generation wireless networks, but optimization for wireless will always be an issue as wired bandwidth increases and mobile applications become more robust. Therefore, even assuming you will be using networks that can offer speeds equal to or faster than dial-up connections, *you should still optimize your application for wireless communication.*

This can be done in a variety of ways. First, data sent over the air should be compressed. Second, unnecessary headers and "handshake" packets should be stripped out of the communication. Finally, the application should be designed to send the minimum amount of data over the air. To cite an extreme example, examine the inefficiency of doing a 3270 terminal emulation session over the air. For each keystroke, the entire screen is being sent over the air. Clearly this is not an application well suited for wireless. Nearly all mobile middleware products offer some level of bandwidth optimization.

Application Portability

An important issue to consider is your migration path and future plans. While you may use one device or network today, you may choose to switch to a different platform in the future. Or, you may even find that different devices work better for different users within your mobile work force. How will you support all these different devices? You don't want to rewrite your application each time you switch out a device because that is time consuming and wasteful of resources. In this instance, you may want to consider using a mobile middleware product that supports many different devices and operating systems to protect your investment in your application.

Ease of Use

Mobile applications must be very easy to use and recover when problems occur. Due to the remote nature of mobile workers, it becomes increasingly difficult to provide troubleshooting and support when problems arise. Applications should be very robust and reliable to work well in the field. Also keep in mind, if you are developing an application for a vertical market, chances are your mobile workers may not be very PC-literate. Therefore applications require an easy-to-use interface with simple and clear screens.

Security

Security is an important concern in the mobile and wireless world. The same level of security you implement within your organization should extend beyond the walls to your mobile users. Consider implementing firewalls as well as security and privacy technologies such as user authentication and data encryption as part of your mobile solution.

Consider implementing firewalls

as well as security and privacy technologies such as user authentication and data encryption

Wireless Awareness

Since your users will be mobile, it is important to provide them with important information that affects the performance and usage of the application. For example, your application should let users know if they are in coverage so they don't waste valuable time and battery power attempting to continually resend messages or connect to the network.

Users also need to know how much battery power they have remaining so that they can make appropriate arrangements. You don't want your users to be caught unaware without power.

Power Management

Just as battery power is an issue to users who need to understand how much power they have in order to plan appropriately, power management is also an issue for the application. Sending and receiving data uses battery power. Therefore, your application should be designed to manage the use of power. For example, it should send the minimal amount of data over the air. It should also have a "sleep" mode during periods of inactivity to minimize the user of power.

Network Management

As mentioned previously, remote users are especially difficult to manage, monitor, troubleshoot, and support. When developing an application, consider how support staff can perform remote diagnostics and manage mobile users. Consider implementing a mobile network management product that will provide alerts and statistics; let your network managers know how many users are connected, how much data is being sent, how many messages are waiting, when users last connected, what network/device they are using, and other such key data.

Remote distribution of software updates is also a key issue for mobile applications. Unless mobile workers regularly come into the office with their computers, you will need to plan for how you will remotely distribute software upgrades to their devices. Software packages exist to help you manage this process, and remote distribution of software is typically done over a wireline dial-up network connection in the evening or morning.

Backward Compatibility

While the issue may appear minor, backward compatibility is particularly important in a mobile environment where not all users' computers will be updated at the same time. You will therefore have servers running with clients of varying versions. If backward compatibility is not ensured, the system will undoubtedly fail.

These are just some key issues that should be considered to allow your application to perform optimally in a mobile environment. If you are implementing a new system for your field force, you may want to consider a packaged application that has already incorporated many mobile-specific functions for your market that you would need to develop, such as mapping, routing, and dispatching, which are not part of your current back-end applications.

Testing

Before starting to pilot or roll out your wireless solution in the field, it first must be put through a rigorous quality assurance testing process. In addition to testing in the lab, it cannot be stressed enough the importance of also testing in the field. For mobile applications more than any, the lab environment is extremely different from the actual environment the application will be used in; no application deployment will be successful off the bat without extensive field-testing. The following leads you through the steps you need to take to develop and execute a quality assurance and testing process for your wireless solution.

Testing Checklist

- ☑ Develop the test plan. This should include what items will be tested, required results, who will be responsible and involved in the testing, and the testing timetable.
- ☑ Build the test bed. The testing environment must be exactly the same as the production environment to ensure accurate results. Therefore, before advancing to this stage, final decisions regarding the hardware, network and software elements must be completed, and the components must be purchased in enough quantity to complete effective testing.
- ☑ Execute testing.
- ☑ Document test results and provide feedback to development. If the test results do not meet the required specifications, the solution must be refined until all requirements are met.
- ☑ If necessary re-test all subsequent revisions until it matches specification.

Developing the Test Plan

To effectively test the system, you must first develop a test bed and test plan. If possible, the majority of the testing should be automated, rather than manual.

A complete test plan should include the following tests/items:

- *Traffic stress testing.* Many applications will perform well with only one user. If you are planning a large-scale rollout, traffic stress testing is increasingly important. In this process you need to test how the application, server, and network handle large volumes of traffic. This testing should also be conducted at various times throughout the day to determine the performance even at "peak" traffic times.

- *Coverage conditions.* Coverage and coverage conditions will vary in the field. In addition to testing for coverage, you should also test to see how your application performs when users move in and out of coverage.
- *Speed.* The solution should also be tested to determine if it is communicating at adequate speeds. If not, what can be done to optimize the communication to improve performance? If speeds are poor and the application is experiencing many disconnections from the network, you may want to consider adjusting the timers that are set within the application determining when the connection is deemed a failure and dropped.
- *Soak testing.* Perform long term tests of usual traffic loads to determine the performance over a period of time.
- *Making sure that "read" is keeping up with "write."* By this, you should determine if the application is processing data quickly enough to work effectively.
- *System failure testing.* The solution should be tested to see how it would react if a component of the system failed or went down. This testing should be performed for each element of the component including the network, server, application server, mobile device, modem, battery, etc.
- *Error testing.* All possible error scenarios should be tested to see the effect on the solution. This would include everything from user error to system crashes. Testing should include not only what happens, but also how the application handles different errors.
- *Testing of all possible functions and uses of the application and system.* Each feature that has been built into the system needs to be tested under each scenario. This would include things like testing to see if messages are stored when the user is out of coverage (if that feature is offered).

- *Message size.* How does the solution handle large messages? Does message size affect the performance? If so, should elements be adjusted to make the application work more effectively?

- *Network switching effects.* If you plan on using multiple networks, be sure to test to see how well the solution responds to network switching. This should be tested under many different scenarios to determine the consistency of the solution.

- *Performance statistics.* Performance statistics should also be documented and compared to the solution requirements to determine if the solution is meeting those requirements.

- *Test of simulators developed in test bed.* If simulators were used in the test bed, these also should be tested to ensure they are working properly and adequately portraying real-life scenarios.

The test plan also needs to provide a mechanism for tracking, documenting, and reporting test results and comparing these results with the required specifications. If the results do not meet the specifications, this must be fed back to the system architect and application developers so that they can improve the performance. Each new revision of the product must then go through the same rigid test procedures to provide quality assurance.

Creating a Test Bed

The test bed environment must be virtually identical with the production environment to successfully intercept potential problems and to be as realistic to actual use as possible. The testing environment should include all required infrastructure and connections to the network(s). As stated earlier, the test bed should be created in the lab *and* in the field.

Execute Testing

Conduct complete testing on each of the elements in the test plan. If you find problems report them to the project team. Once the next version is created, the solution must be retested from the beginning. Go through the entire testing process with each product because it is impossible to ensure that one fix will not adversely affect another element of the solution. This process must be continued until all requirements are met. In the future, as the solution is enhanced or upgraded, repeat this process until it can be assured that the solution is working properly and performing according to requirements.

Staging

Before deploying your solution, you need to first "stage" the equipment in preparation for rollout. This step should be taken prior to the pilot stage and continue with each new device that is deployed.

The steps required of this stage are outlined below.

1. Loading of software onto the mobile device.
2. Configuring the device and software for the individual user. This includes setting up the user name and password.
3. Registering the modems with the wireless network provider.
4. Testing of each device to ensure that everything is loaded and working according to specification.
5. Asset management. Each device should be tracked and managed so that it is easily determined who is using what types of equipment and where the device resides.
6. Training. As mentioned in the other section, each user should be trained on the operation of the device, application, and modem prior to sending them out into the field.

Piloting Your Application

Your primary consideration when piloting your application is to test and see if the new way of doing business works. Therefore, you need to get as representative a sample as possible and include training and feedback mechanisms for users. Ideally, the system will run in parallel to existing systems so that if problems are encountered, it is easy to cut back to the existing system. A proper support system is absolutely essential for a successful pilot. The pilot phase is important not only in testing results of the new system, but also in generating user buy-in. If the pilot users like the new system, they will become your champions, and you'll have a much easier time getting all users excited about using the system.

Developing the Pilot Plan

The first step in a successful pilot is developing a comprehensive pilot plan.

These steps are outlined below.

1. Identify your pilot users. As stated, this should consist of a representative sample from each area or job role.

2. Determine what you want to get out of the pilot. For example, are there specific results that need to be documented to determine ROI or effectiveness? Is additional performance testing required? Through this analysis you will be able to determine what you want to test during the pilot phase.

3. Decide upon the necessary length of the pilot. To show true representation, all pilots should be a minimum of 30 days.

4. Train users on what the solution is supposed to accomplish, new business processes, the application, device operation, troubleshooting, who to call in times of difficulty, and feedback requirements.

5. Slowly roll out to pilot users. Anticipate a lot of hand holding in the first week as users get used to the new system. You may even want to consider having someone from the project team riding along with the pilot users to assist them if they are confused or experience difficulties.

6. Gather feedback. Be sure to provide your users with feedback guidelines or formats such as reports, surveys, focus groups or interviews.

7. Adjust the system as required and continue to full-fledged rollout efforts.

Ideally,
the system will
run
in parallel to
existing systems

Pilot Testing

During the pilot phase, you should test the performance as well as the results generated by the system. This will determine not only if the system is designed correctly and usable in the field, but also if it has affected the business process to such a degree as to generate a competitive advantage. Without showing proven business results, the project will not move beyond the pilot phase.

The following list outlines some items to test:

- Performance Testing
  - Scalability
  - Usability
  - Application usage with network coverage
  - Coverage
  - Radio awareness
  - Wireless performance
- Business Testing
  - Time saved
  - Number of visits/calls/jobs conducted compared to prior results
  - Ability to generate additional revenue (e.g., through selling warranties on the spot)
  - Quantifiable improvements in customer service (may be tested through before and after customer surveys)
  - Jobs or tasks eliminated
  - Job safety improvements

Once you've convinced the project team and company that the solution is operating smoothly and have demonstrated positive results, it is time to move on to complete roll-out.

Rolling Out

When fully deploying your application, it is often best to do so in stages to groups of users. Each stage will include training for users, distribution of equipment and mechanisms to provide feedback, and measure results. These are similar steps to those followed when performing the pilot testing, but on a much broader scale. At this stage, it is absolutely critical to have all your ducks in a row including documentation, training, and support.

Finding the Right Partner

Developing and deploying a mobile solution is a major project that requires buy-in and commitment from many areas of the organization if it is to be successful. While it may seem like an intimidating task for those unfamiliar with wireless data communication, there are many experts available to help guide you through it and assist you with the design, development, and deployment of your solution. A successful deployment is much easier achieved with the backing of strong partners with a deep understanding of the wireless industry.

For example, services provided by a partner include:

- Analyze business processes and define new streamlined, mobile methods
- Create business justification document
- Prepare RFPs
- Review RFP responses
- Evaluate software, hardware, or network

- Prepare specification/requirements documents
- Produce coverage analysis
- Review application design
- Develop application
- Integrate application with backend systems
- Develop application agents
- Review wireless test plan
- Design and construct wireless test environment
- Perform wireless testing
- Develop deployment plan
- Set up support environment
- Support system troubleshooting and fine-tuning
- Train developers on middleware
- Train users on new applications

Conclusion

In many respects, wireless application development is exactly the same as developing applications for landline networks. Yet, the differences in wireless networks, mobile devices, and the mobile user's experience are significant enough to place additional burdens on the system architect and application developer.

Given the early market nature of wireless data, corporations should amass a team of business and IT talent from within the company and wireless experts from outside the company to design new business processes and the wireless applications that will make them successful wireless businesses.

The next chapter deals with the strategic questions surrounding wireless data and where within the corporation this disruptive technology can deliver substantial business improvements.

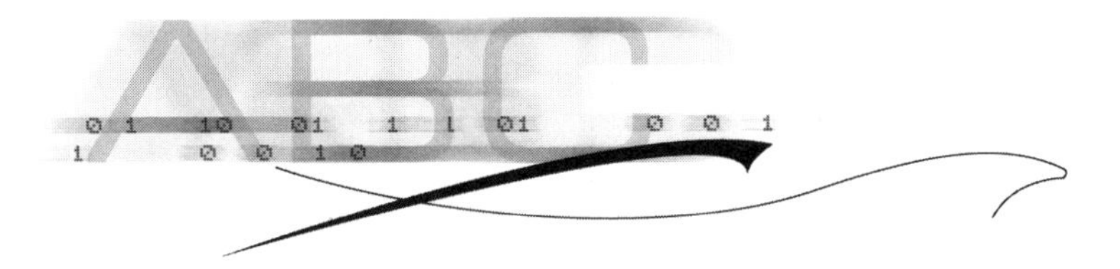

9 BOLDLY GO...INVENTING THE WIRELESS ENTERPRISE

LIKE SCHOOL CHILDREN IN THE PLAYGROUND ROCKING rhythmically back-and-forth as they time their entrance into the moving jump ropes of "double-Dutch," corporate management sways toward and retreats from wireless data. Managers find themselves attracted by the potential return-on-investment that they can achieve while at the same time they sense risk and potential failure given the current state of the art of the technology. This chapter acknowledges the tension of these two forces and provides a clear set of recommendations of where, when, and how to proceed forward. We encourage the reader to use this chapter to identify the most successful application of wireless data within their enterprise.

In the earlier chapters of this book, several case studies and examples of companies were presented who have deployed wireless data in line-of-business applications. FedEx captures customer signatures, swipes bar code package data and time stamps the movement of packages all in real-time. Their results are clear: improved customer service and decreased costs in delivering packages. Their core business function has been modified and improved by wireless data, providing a clear ROI and a visible success within the company.

Common Knowledge Doesn't Always Apply

Not every company has this opportunity to change their business so dramatically. Or is this the underlying error of what is known as "common knowledge?" Most managers, both technical and functional, look around at their peers and read the headlines of their respective trade journals. Obviously at the introductory phase of a technology, with few deployments underway, your peers have not undertaken such a project. Likewise, there are few headlines, given that only the early companies have embarked on a wireless data journey. So the "common knowledge" becomes, "it is too early for me to consider how wireless data can change my business processes." Yet, this is the WRONG conclusion.

A quick illustration will emphasize this point. Recently, we did a workshop where over a dozen major corporations were represented. As the group discussed the ability of wireless data to lower costs within a corporation, to raise revenues, or to reach new markets, one member of the group openly disagreed. "Well it's clear that FedEx could achieve these benefits, but my business doesn't involve package delivery so I don't see the relevance of wireless data." He worked at an insurance company and was confident in his position that it would be years before he would or could take advantage of this technology.

Many participants in the workshop were beginning to show doubt themselves as the statement lingered in the air. Is it true? Is wireless data so narrowly defined as a technology suited only to solve the problems of package tracking? No.

The evidence is clear that companies and government agencies around the world have creatively used wireless and mobile data to improve their operations and grow their businesses. Insurance companies, like the one the "doubting Thomas" is from, process claims *in the field* for their customers without having to go back to the home office and complete paperwork. Customers are ecstatic about the speed of the "wireless field process," and the insurance companies are reducing their costs to process a claim, which are both key indicators of business performance.

The "doubting Thomas" was unaware of his competitors' actions (yes, there are many more than one insurance company doing field claims processing in real-time) and quite surprised to learn that some had begun their projects over two years ago. Although he was embarrassed, he was not alone in believing that "no one was actively running their business on wireless data." Few in the room were knowledgeable of the course and speed at which companies are using the technology.

Determining Whether or Not to Go Wireless

Consider these real-life, line of business applications

- A client has just looked at a center hall colonial in the $400,000 range and likes the neighborhood. They ask, "are there other homes like this within walking distance of the high school?" Today, some real estate agents are able to request this information through a PDA or laptop in their vehicles. In minutes, the agent has all of the information they need to show the client every house that meets their criteria.

- Fixing appliances on location can be a profitable business if you limit the number of return trips required to complete a repair. From small companies that repair heating, ventilating, and air conditioning equipment (HVAC) to large fleets of white-goods repair personnel that do their work in people's homes, wireless data improves scheduling, parts ordering, and billing.

- At large banks, personal banking is a very profitable and competitive part of the business. Skilled members of the personal banking team meet with their clients who are known as "high net worth individuals" – individuals who have large amounts of disposable income. To maintain their critically important client relationships, some banks are using wireless data to advise their clients anywhere and in real-time. This is particularly important because the bank's clients demand timely information independent of where they may happen to be when they need it. The banks that offer this level of service are differentiating themselves, resulting in higher customer retention rates and winning new customers.

- Architects are using wireless data to compare on-site conditions with the latest changes being made at the central office, resulting in less time, less effort, and lower rework on the site.

- Police officers are using PDAs and laptops to send and receive information regarding vehicles and individuals that assist in arrests.

- Home health care nurses are "filing their paperwork" as they move from one patient to the next. No longer do they have to get out of the field and spend hours each day completing their files. Their employers are also finding a side benefit – it is easier to retain these nurses because their jobs are more focused on patient care than on paperwork. In light of a critical shortage of nurses, this is a powerful advantage.

Where Wireless Fits

Beyond providing substance to the argument that wireless data is improving businesses, this wide range of examples offers insight into patterns of where wireless data fits in a company. The factors are tri-fold.

First, a customer or an employee must have a *need*. Although this could be overlooked as so apparent as not to need mention, identifying the need is the core reason for any application. Remember "desire" from Chapter 5.

Second, there must an *element of mobility* where the user is not at her desk for a reasonable portion of the business day.

Third, there must be some *time urgency* to the data while the user is mobile. This can be data that the user wants to send to the corporation (package tracking information so others can look at this data in near real-time) or data the corporation wants to send to the user (such as dispatch information).

Wireless data, being a disruptive technology, brings a new twist to the technology landscape – freedom of location. This single parameter of difference enabled each of the applications listed earlier. Prior to being able to deliver and send data while on the move, all of the earlier scenarios were impossible to effect.

Factors Influencing the Wireless Data Equation

Looking at your business, when are workers (or suppliers, or distributors, or customers) *mobile* and in *need* of *time-sensitive data*? This is the question that will define the extent to which your business can possibly benefit from wireless data. The nexus of these three factors defines the necessity for wireless data technology.

Interestingly, the factors are additive, meaning that more of one of them, such as more mobile workers, increases the potential of business value for wireless data. Numerically, if you have 10 workers who are mobile out of 10,000 employees, you have less potential for improving your business than if 1,000 of your workers were mobile. Likewise, if your mobile employees are mobile for all 8 hours of their work day, your potential for business advancements is greater than if your workers only spend one hour of their working day out of the office.

The other two factors have similar effects on the wireless data equation. A higher need and greater time criticality of the data enhances the potential return of deploying the technology. Repair teams that fix a broken industrial compressor system have a high degree of need for time sensitive data. Every moment that they are delayed in placing it back on-line increases the cost of repair to their company and may also place in jeopardy their ability to meet their Service Level Agreement with their customer. The breach of the service agreement may have penalties and other costs that add to the potential value of using wireless data to inform the repair team of how to do problem determination and take corrective action.

Real Life Calculations

Returning to the insurance executive who didn't see the need or potential for using wireless data in his business, a quick analysis changed his opinion.

Specifically:

- He estimated that over 400 employees were mobile in his division
- Of these 400 employees, they were "outdoors about 50% of their day," meaning the equivalent of 200 full time employees
- In performing their work, he identified one process that required time-sensitive data, such as knowing the customer's payment history and sending collected data in from the field. He also realized that he didn't have a good feel for the daily routine of these employees and readily admitted that he would have to learn how much data they could use and how much of it was time critical.

His last point is possibly the most telling in that mobile workers represent a superb opportunity for data technology simply, if for no other reason, their jobs have not been automated by technology because it was not possible until now. Consider the richness of the opportunity in this particular scenario. A 200-person workforce that has not been touched by automation and who work in front of customers will generate a significant return on investment!

Is Wireless Right for *Your* Business?

Does your business have the proper mixture of mobility and the need for time-sensitive data? Upon inspection, most companies find that they have overlooked this workforce and the potential improvements that can be made in their

processes. Some companies have provided laptops, but they are not wirelessly enabled and provide no means for delivering real-time, location sensitive data. More importantly, these workers suffer from being "out of site, out of mind." Their business processes are less well known to others in their company. Basically, this workforce is a flashing billboard that shouts out "HELP ME IF YOU CAN."

Having described the three factors of need, mobility, and time-sensitivity, we believe it is helpful to review three different situations that can arise once the numerical business case is evaluated. In other words, once you have determined (at least by doing a back of the envelope calculation) that there is potential merit in further studying how to change your business with wireless data, you will come to a fork in the road where there are three choices.

Three Paths to Wireless Enabling Your Business

1) *The business opportunity has a clear ROI measured in hard dollars, which compels you to move forward.* The application affects your department and you have control over the funds and the resources to plan, execute and measure its success. This is the best of all the possibilities since you control your destiny on this project and with high certainty, you will able to deliver tangible business results. The previous chapter provides the details necessary to get started. Many business processes fall into this category and are waiting to be discovered by a functional or information technology manager to drive by the flashing billboard.

2) *The business opportunity has a clear ROI but across multiple departments, business processes, or divisions.* This situation is common especially when the number of mobile workers is not a significant enough factor or when the costs of implementing the project simply cannot be justified by a single application. Flashback to the 1970s, when the first department (accounting) tried to justify a mainframe with one application (payroll). Although the mainframe application would clearly be better than doing the business process manually, all of the startup costs of the project, such as training the personnel, building a raised floor room, putting in water cooling, et cetera, overwhelmed the benefits. Businesses did not walk away from the opportunity, rather they found

other business processes that could benefit from the same general investment in the mainframe and by sharing costs; each project became cost justifiable. This is sometimes the case for business processes that can be enabled by wireless data. In these situations, the focus needs to be on finding a second application within another department or division that can share the costs of the project. Typically, cost sharing can be achieved on the middleware since this is a reusable platform for multiple applications. Other components that can be reused are the devices themselves (if the same target user group will use both applications); user and IT training; airtime and telecommunication base charges, and server backend integration software charges.

Choose a strong business manager to act as the overall champion responsible for the outcome of the combined project. This is mandated if the project spans more than one department or division.

3) *The business opportunity has no clear ROI, but there is a strong feeling that it should be undertaken.* Projects of this nature can be successful, however, be cautious of naysayers within the company who might pounce on an undertaking that failed to deliver on its promises. We recommend projects like these be initiated with a set of objectives that includes "lessons learned" and are termed "pilots" so as to set the proper expectations. If you have 100 mobile professionals, choose a representative sample of this group (10 to 20) and deploy the application with them. Follow them around and learn how they use the application and how it changes the way that they perform their work. Use this as an opportunity to explore other applications and business processes that could be improved. Generally, the user population has a wealth of ideas once they are exposed to the technology and begin to think wirelessly.

The doubting Thomas insurance executive has returned to his office and has begun his evaluation to determine which of the three forks in the road he can take. What is eminently clear to him now is that there are three options where he previously contended that he was not traveling down the road towards a wireless business. He is not alone in his incorrect assumption that wireless data cannot benefit companies today. Most managers are held by the inertia of the common knowledge we spoke of at the beginning of the chapter where everyone looks around to find others looking back at them, wondering when will be the right time to make a move on wireless data.

The Time is Now

Andy Grove of Intel lives by the motto "only the paranoid survive." Given the evidence that companies around the world are using wireless data to reach new customers, lower their costs of delivering services, and grow their revenues, managers can no longer believe that their peers are not using wireless data. In industries where the workforce is naturally mobile, such as in less than truckload hauling, a recent survey of 150 companies showed that almost half of them use wireless data to improve pickup and delivery dispatching. Like the insurance executive's surprise that his competitors were already using wireless data to assist field workers, you have to be wondering "am I going to be the leader or am I already a follower?"

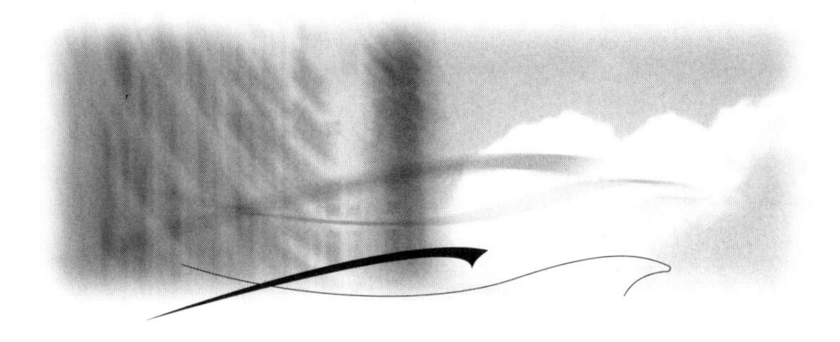

EPILOGUE

As a disruptive technology, wireless data will move over time from the periphery of the information technology space to its core where it will be a full-time complement to the wireline world that we all know. In that future, our "visual knowledge" of the landscape around us will be significantly improved as we will be able to sense, in real-time, the price changes at gas pumps as we drive down the street and our fuel tank dips below one quarter of a tank.

Many, many more applications will be designed in the next few years that we cannot today begin to imagine. It's as if paint, brush and canvas have been created, and the schools of painting have yet to show us their creative genius and uniqueness. With the availability of wireless application platforms, wireless devices, and networks, the tools are now ready for the skilled artisans to paint masterpieces of all kinds. This is truly a dawning of a new era in computing.

As we watch the artwork of wireless data transform work processes in government and businesses around the world, we have begun to collect the ideas in a living workspace. We have elected to dedicate a portion of our web site to aggregate the collective energy that is being applied to the use of wireless data in the enterprise. The site is for your education, amusement, and input. The focus is on you and helping you make a sound decision. The site will contain applications across all industries and geographies, business cases that were used to justify these applications, tips and tricks, reference accounts, conference listings, and other items that grow out of a desire to inform the business community on the value of wireless data. The site can be found at **www.broadbeam.com.**

We look forward to you joining us at the web site and providing us feedback on the contents of this book and the speed and direction of your company's use of wireless data.

GLOSSARY

The following glossary defines some of the most common terms used in wireless communications, networking and telecommunications to allow you to become more fluent in the vocabulary of vendors and suppliers.

Advanced Radio Data Information System (ARDIS)

Packet-switched wireless network that delivers a data rate of 19.2 Kbps and provides deep penetration into buildings. Primarily used for field service and transportation applications, ARDIS was created by Motorola in the mid 1980s for IBM's field service division. It was later spun off as a commercial service.

Analog Mobile Phone System (AMPS) Network

The analog cellular mobile phone system that serves North and South America, as well as over 35 other countries. It uses FDMA (Frequency Division Multiple Access) transmission in the 800 Mhz band.

Application Program Interface (API)

A language and message format used by an application program to communicate with an operating system, or some other system or control program such as a database management system (DBMS) or communications protocol.

Bandwidth

Bandwidth (the width of a band of electromagnetic frequencies) is used to mean (1) how fast data flows on a given transmission path, and (2) somewhat more technically, the width of the range of frequencies that an electronic signal occupies on a given transmission medium. Any digital or analog signal has a bandwidth. Generally speaking, bandwidth is directly proportional to the amount of data transmitted or received per unit time. In wireless networks, bandwidth refers to the range of frequencies occupied by a modulated radio-frequency signal, usually given in hertz (cycles per second) or as a percentage of the radio frequency. For example, an AM (amplitude modulation) broadcasting station operating at 1,000,000 hertz has a bandwidth of 10,000 hertz, or 1 percent (10,000/1,000,000). The term also designates the frequency range that an electronic device, such as an amplifier or filter, will transmit.

Bluetooth

A short-range RF standard for wireless communication between devices which permits transmission speeds of up to 1Mb between devices that are within a 30-foot range of one another. Also known as a "cableless" connection to replace a myriad of cables linking wireless PDAs, phones, laptops, and other devices.

C-Band Communications

Communications transmitted via satellite with an uplink frequency at 6GHz and a downlink frequency at 4GHz.

Cellular Digital Packet Data (CDPD)

Digital wireless transmission system deployed as an enhancement to an existing analog cellular network. Based on IBM's CelluPlan II, it provides a packet overlay onto the AMPS network and moves data at 19.2 Kbps over ever-changing unused intervals in the voice channels. CDPD is used for applications such as public safety, point of sale, mobile positioning, and other business services.

Circuit-Switching

A networking technology that provides a temporary, but dedicated, connection between two stations no matter how many switching devices the data is routed through. Circuit switching was originally developed for the analog-based telephone system in order to guarantee steady, consistent service for two people engaged in a phone conversation. Analog circuit switching (FDM) has given way to digital circuit switching (TDM), and the digital counterpart still maintains the connection until broken (one side hangs up). This means bandwidth is continuously reserved and "silence is transmitted" just the same as digital audio.

Code Division Multpile Access (CDMA)

A digital wireless technology that uses a spread spectrum technique, which spreads a signal across wide-frequency bands. CDMA offers increased system capacity, increased voice quality, fewer dropped calls, and IP data services. Signals are either selected or rejected at the receiver by recognition of a user-specific signature waveform, which is constructed from an assigned spreading code.

cdmaOne™

A brand name, trademarked and reserved for the exclusive use of CDMA Development Group member companies, that describes a complete wireless system that incorporates the IS-95 CDMA air interface, the ANSI-41 network standard for switch interconnection and many other standards that make up a complete wireless system.

CDMA2000

CDMA2000 is a name identifying the third generation technology that is an evolutionary outgrowth of cdmaOne offering operators who have deployed a second generation cdmaOne system a seamless migration path that economically supports upgrades to 3G features and services within existing spectrum allocations for both cellular and PCS operators. CDMA2000 supports the second generation network aspect of all existing operators regardless of technology (cdmaOne, IS-136 TDMA, or GSM). This standard is also known by its ITU name IMT-CDMA Multi-Carrier (1X/3X).

Competitive Local Exchange Carrier (CLEC)

A non-RBOC organization that offers local telephone services. Although most CLECs are established as a telecommunications service organization, any large company, university, or municipal government has the option of becoming a CLEC and supplying its own staff with dial tone at reduced cost. It must have a telephone switch, satisfy state regulations, pay significant filing fees, and also make its services available to outside customers.

Customer Resource Management (CRM) System

Legacy systems with wireless links today that enable sales and marketing personnel to manage contacts and customer lists, current contracts, etcetera.

Database Management System (DBMS)

A software system that facilitates (a) the creation and maintenance of a database or databases, and (b) the execution of computer programs using the database or databases.

Digital-Advanced Mobile Phone Service (D-AMPS)

D-AMPS, sometimes spelled DAMPS, is a digital version of Advanced Mobile Phone Service, the original analog standard for cellular telephone service in the United States. Both D-AMPS and AMPS are now used in many countries. D-AMPS adds time division multiple access to AMPS to get three channels for each AMPS channel, tripling the number of calls that can be handled on a channel. D-AMPS is Interim Standard-136 from the Electronics Industries Association/ Telecommunication Industries Association (EIA/TIA).

Digital Enhanced Cordless Telecommunications (DECT)

A cordless phone standard that is based on TDMA operating in the 1.8 and 1.9 GHz bands. DECT uses Dynamic Channel Selection/Dynamic Channel Allocation (DCS/DCA) to enable multiple DECT users to coexist on the same frequency.

Digital Subscriber Line (DSL)

A technology for bringing high-bandwidth information to homes and small businesses over ordinary copper telephone lines. xDSL refers to different variations of DSL, such as ADSL, HDSL, and RADSL. Assuming your home or small business is close enough to a telephone company central office that offers DSL service, you may be able to receive data at rates up to 6.1 megabits (millions of bits) per second (of a theoretical 8.448 megabits per second), enabling continuous transmission of motion video, audio, and even 3-D effects.

E-commerce

Business relationships and selling information, services, and commodities are maintained by means of computer telecommunications networks.

Elliptical Curve Cryptography (ECC)

A public key cryptography method that provides fast decryption and digital signature processing.

Enhanced Digital Access Communications System (EDACS)

A private radio or specialized mobile radio network designed by Ericsson Inc. EDACS is mostly used in public safety applications to ensure in coverage and bandwidth are always available (since public networks could be clogged since they don't control the channels). Proprietary control channel that allows emergency signals to be instantly displayed at the control center with officer identification, last location, and activity information displayed.

Enhanced Data rates for Global Evolution (EDGE)

An enhancement to the GSM and TDMA wireless communications systems that increases data throughput to 384 Kbps.

Enterprise Resource Planning (ERP)

ERP is an industry term for the broad set of activities supported by multi-module application software that helps a manufacturer or other business manage the important parts of its business, including product planning, parts purchasing, maintaining inventories, interacting with suppliers, providing customer service, and tracking orders. ERP can also include application modules for the finance and human resources aspects of a business.

EPOC

A 32-bit operating system for handheld devices with wireless access to phone and other information services from Symbian Ltd., London, UK.

Extranet

A private network that uses IP communication over the public telecommunication system which a company extends to conduct business with its suppliers, vendors, partners, customers, or other businesses.

Federal Communications Commission (FCC)

The U.S. government agency that regulates interstate and international communications including wire, cable, radio, TV, and satellite.

FLEX

A Motorola Inc.-licensed protocol that gives carriers more capacity on their networks and faster transmission times. Also refers to the FLEX family of protocols – FLEX, InFLEXion and ReFLEX.

Frequency Division Duplexing (FDD)

A transmission method that separates the transmitting and receiving channels with a guard band.

Frequency Division Multiple Access (FDMA)

The division of the frequency band allocated for wireless cellular telephone communication into 30 channels, each of which can carry a voice conversation or, with digital service, carry digital data. FDMA is a basic technology in AMPS networks, the most widely-installed cellular phone system installed in North America. With FDMA, each channel can be assigned to only one user at a time. FDMA is also used in the Total Access Communication System (TACS).

Frequency Hopping Spread Spectrum (FHSS)

Frequency hopping is one of two basic modulation techniques used in spread spectrum signal transmission. It is the repeated switching of frequencies during radio transmission, often to minimize the effectiveness of "electronic warfare" – that is, the unauthorized interception or jamming of telecommunications. In an FHSS system, a transmitter "hops" between available frequencies according to a specified algorithm, which can be either random or preplanned. The transmitter operates in synchronization with a receiver, which remains tuned to the same center frequency as the transmitter. A short burst of data is transmitted on a narrowband. Then, the transmitter tunes to another frequency and transmits again. The receiver thus is capable of hopping its frequency over a given bandwidth several times a second, transmitting on one frequency for a certain period of time, then hopping to another frequency and transmitting again. Frequency hopping requires a much wider bandwidth than is needed to transmit the same information using only one carrier frequency.

General Packet Radio Service (GPRS)

An enhancement to the GSM mobile communications system that supports data packets. GPRS enables continuous flows of IP data packets over the system for such applications as Web browsing and file transfer. GPRS differs from GSM's short messaging service (GSM-SMS), which is limited to messages of 160 bytes in length.

Global Positioning Systems (GPS)

This is the only system that is able to show you your exact position on the Earth anytime, in any weather, anywhere. GPS satellites, 24 in all, orbit at 11,000 nautical miles above the Earth. They are continuously monitored by ground stations located worldwide. The satellites transmit signals that can be detected by anyone with a GPS receiver. Using the receiver, you can determine your location with great precision.

Global System for Mobile communication (GSM)

GSM is a digital mobile telephone system that is widely used in Europe and other parts of the world. GSM uses a variation of time division multiple access and is the most widely used of the three digital wireless telephone technologies (TDMA, GSM, and CDMA). GSM digitizes and compresses data, then sends it down a channel with two other streams of user data, each in its own time slot. It operates at either the 900 MHz or 1800 MHz frequency band.

Graphical User Interface (GUI)

A graphics-based user interface that incorporates icons, pull-down menus, and a mouse. The GUI has become the standard method of interacting with a computer. The three major GUIs are Windows, Macintosh, and Motif.

Handheld Device Markup Language (HDML)

Often compared to Wireless Markup Language (WML), HDML is a language that allows the text portions of Web pages to be presented on cellular telephone and personal digital assistants via wireless access. Developed by Unwired Planet, HDML is an open language offered royalty-free.

HyperText Markup Language (HTML)

The document format used on the World Wide Web. Web pages are built with HTML tags or codes embedded in the text. HTML defines the page layout, fonts, and graphic elements as well as the hypertext links to other documents on the Web.

HyperText Transfer Protocol (HTTP)

The set of rules for exchanging files (text, graphic images, sound, video, and other multimedia files) on the World Wide Web. Relative to the TCP/IP suite of protocols (which are the basis for information exchange on the Internet), HTTP is an application protocol.

Institute of Electrical and Electronics Engineers (IEEE)

The IEEE fosters the development of standards that often become national and international standards. The organization publishes a number of journals, and has many local chapters.

IIEEE 802.11

In wireless LAN (WLAN) technology, 802.11 refers to a family of specifications developed by a working group of the Institute of Electrical and Electronics Engineers (IEEE). There are three specifications in the family: 802.11, 802.11a, and 802.11b. The 802.11 and 802.11b specifications apply to wireless Ethernet LANs, and operate at frequencies in the 2.4-GHz region of the radio spectrum. Data speeds are generally 1 Mbps or 2 Mbps for 802.11, and 5.5 Mbps or 11 Mbps for 802.11b. The 802.11b standard is backward compatible with 802.11.

i-Mode

A packet-based information service for mobile phones from NTT DoCoMo (Japan). i-mode provides Web browsing, e-mail, calendar, chat, games, and customized news.

IMT 2000

A framework from the ITU for third generation (3G) world-wide wireless phone standards that delivers high-speed multimedia data as well as voice communications.

Independent Software Vendor (ISV)

An individual or business that develops software. ISV refers to a vendor that specializes in software only, and is not part of a computer systems or hardware manufacturer.

Internet

A network that connects many computer networks and that is based on a common addressing system and communications protocol called TCP/IP.

Internet Protocol (IP)

IP is a network-layer protocol that contains addressing information and some control information that enables packets to be routed. IP is the primary network-layer protocol in the Internet protocol suite. Along with the Transmission Control Protocol (TCP), IP represents the heart of the Internet protocols. IP has two primary responsibilities – 1) providing connectionless, best-effort delivery of datagrams through an internet work; and 2) providing fragmentation and reassembly of datagrams to support data links with different maximum-transmission unit (MTU) sizes.

Intranet

Using common Internet protocols or core technologies in conjunction with their own business applications, corporations can easily communicate, distribute information, and facilitate project collaboration across the entire enterprise, while keeping unauthorized users out.

Integrated Services Digital Network (ISDN)

An international telecommunications standard for providing a 128 Kbps digital service from the customer's premises to the dial-up telephone network.

International Telecommunications Union (ITU)

Globally oriented standards body based in Geneva, Switzerland, that promulgates wireless and other telecommunications standards.

IS-136

The second generation of the TDMA digital cellular system operates in North America in the 800 MHz band and 1.9 GHz PCS band.

Java

Java is an object-oriented programming language, created by Sun Microsystems, Inc., expressly designed for use in the distributed environment of the Internet. Java's great promise was of "platform independence" in which it allows software developers to generate applications that can run on all hardware platforms, small, medium and large, without modification. A Java source program is compiled into what Java calls "bytecode", which can be run anywhere in a network on a server or client that has a Java virtual machine. The Java virtual machine interprets the bytecode into code that will run on the real computer hardware.

Java 2 Platform, Enterprise Edition (J2EE)

A Java platform designed for the mainframe-scale computing typical of large enterprises. Sun Microsystems (together with industry partners such as IBM) designed J2EE to simplify application development in a thin client tiered environment. J2EE simplifies application development and decreases the need for programming and programmer training by creating standardized, reusable modular components and by enabling the tier to handle many aspects of programming automatically.

Java 2 Platform, Micro Edition (J2ME)

A technology that allows programmers to use the Java programming language and related tools to develop programs for mobile wireless information devices such as cellular phones and personal digital assistants (PDAs). J2ME consists of programming specifications and a special virtual machine, the K Virtual Machine, that allows a J2ME-encoded program to run in the mobile device.

Liquid Crystal Display (LCD)

A display technology that uses liquid crystal sealed between two layers of glass. It is comprised of numerous dots that are charged or uncharged, and that either reflect or do not reflect light, in order to form characters, letters, and numbers. LCDs require minimal power and have fast reaction.

Local Area Network (LAN)

LANs connect computers and peripheral devices in a limited area such as a business office or college campus by means of wires, cables and fiber optics that transmit data rapidly. A typical LAN consists of two or more PCs, printers, and high-capacity disk-storage devices called file servers, which enable each computer on the network to access a common set of files.

Local Multipoint Distribution Service (LMDS)

A digital wireless transmission system that works in the 28 GHz range in the U.S. and 24-40 GHz overseas. LMDS requires line of sight between transmitter and receiving antenna. The distance between the two can be from one to four miles, depending on weather conditions. LMDS provides bandwidth in the OC-1 to OC-12 range, which is considerably greater than other broadband wireless services.

Low Earth Orbit Satellite (LEO)

An orbit plane several hundred feet above the earth into which communication, or LEO, satellites are increasingly being launched. "Big" and "little" LEO satellites are assigned specific radio frequencies. Big LEO satellites support voice and data communications, while little LEO satellites support data only.

Middleware

Software that functions as a conversion or translation layer, consolidator and integrator. Customized middleware solutions have been developed for decades to enable an application to communicate with another that runs on a different platform, or that comes from a different vendor or both.

M-Commerce

M-commerce (mobile commerce) is the buying and selling of goods and services through wireless handheld devices such as cellular telephone and personal digital assistants. Known as next-generation e-commerce, m-commerce enables users to access the Internet without needing to find a place to plug in.

Mobitex

Two-way cellular, radio-based packet-switched data communications system originally developed by Ericsson; implemented in North America by Cingular Interactive (formerly BellSouth Wireless Data) as a ubiquitous wireless data network.

Multicasting

The communication between a single sender and multiple receivers on a network.

Multichannel Multipoint Distribution Service or Microwave Multipoint Distribution Service (MMDS)

A digital wireless transmission system that works in the 2.2-2.4 GHz range. MMDS requires line of sight between transmitter and receiver, which can be 30 or more miles apart.

Open Database Connectivity (ODBC)

An open standard application programming interface (API) for accessing a database.

Orthogonal Frequency-Division Multiplexing (OFDM)

Modulation technique used for digital TV in Europe, Japan and Australia. OFDM was first promoted in the early 1990s as a wireless LAN technology. Its spread spectrum technique distributes the data over a large number of carriers that are spaced apart at precise frequencies.

Packet

A packet is the unit of data that is routed between an origin and a destination on the Internet or any other packet-switched network. When any file (e.g., e-mail message or HTML file) is sent from one place to another on the Internet, the TCP layer of TCP/IP divides the file into "chunks" of an efficient size for routing. Each of these packets is separately numbered and includes the Internet address of the destination. The individual packets for a given file may travel different routes through the Internet. When they have all arrived, they are reassembled into the original file (by the TCP layer at the receiving end).

Packet Switching

A networking technology in which relatively small units of data called packets are routed through a network based on the destination address contained within each packet. This type of communication between sender and receiver is known as connectionless (rather than dedicated). Consequently, all packets in a single message do not have to travel the same path. They can be dynamically routed over the network as circuits become available or unavailable. The destination computer reassembles the packets back into their proper sequence.

Personal Area Network (PAN)

A personal area network is a technology that could enable wearable computer devices to communicate with other nearby computers and exchange digital information using the electrical conductivity of the human body as a data network.

Personal Communications Services (PCS)

Wireless services that emerged after the U.S. Government auctioned commercial licenses in 1994 and 1995. This radio spectrum in the 1.8-2 GHz range is typically used for digital cellular transmission that competes with analog and digital services in the 800 Mhz and 900 MHz bands.

Personal Computer Memory Card International Association (PCMCIA)

This international standards body and trade association was founded in 1989 to establish a standard for connecting peripherals to portable computers.

Personal Digital Assistant (PDA)

PDA is a term for any small mobile hand-held device that provides computing and information storage and retrieval capabilities for personal or business use, often for keeping schedule calendars and address book information handy.

Personal Information Management (PIM)

The digital management of personal information, such as addresses, phone numbers, and calendaring, functions previously managed manually in day planners.

Portal

A term, generally synonymous with gateway, for a World Wide Web site that is or proposes to be a major starting site for users when they get connected to the Web. Portals typically provide a range of services that include Web searching, news, telephone directories, free e-mail, discussion groups, online shopping, and links to other sites.

Public Switched Telephone Network (PSTN)

The worldwide circuit-switched voice telephone network, which is the heart of most telephone networks. Once an analog-only system, it is virtually all digital today.

QWERTY

The standard English language typewriter keyboard. Q, W, E, R, T, and Y are the letters on the top left, alphabetic row.

Random Access Memory (RAM)

The place in a computer where the operating system, application programs, and data in current use are kept so that they can be quickly reached by the computer's process. RAM is much faster to read from and write to than the other kinds of storage in a computer, the hard disk, floppy disk, and CD-ROM. However, the data in RAM stays there only as long as your computer is running. When you turn the computer off, RAM loses its data.

Radio Frequency (RF)

The range of electromagnetic frequencies above the audio range and below visible light. All broadcast transmission, from AM radio to satellites, falls into this range, which is between 30 KHz and 300 GHz.

Radio Frequency Identification Devices (RFID)

Data collection technology that uses electronic tags to store identification data, and a wireless transmission method to capture the data.

Radiotelegraphy

Radio communication that uses Morse Code or other coded signals and where the radio carrier is modulated by changing its amplitude, frequency, or phase in accordance with the Morse dot-dash system or another code.

Regional Bell Operating Company (RBOC)

The regional Bell telephone companies that were spun off of AT&T by court order.

RF Residential Gateway

Integrated services interface for home-based utilities and communications networks.

Short Messaging Service (SMS)

A text message service that enables short messages of generally no more than 140-160 characters in length to be sent and transmitted from a wireless phone. SMS is similar to paging. However, SMS messages do not require the mobile phone to be active and within range and will be held for a number of days until the phone is active and within range. SMS is supported by GSM and other mobile communications systems.

Smartphone

The term Smartphone is sometimes used to characterize a wireless telephone set with special computer-enabled features not previously associated with telephones, especially those common to the PDA such as a built-in address book, and wireless browsing of the web and e-mail.

Specialized Mobile Radio (SMR)

Communications services typically used by taxicabs, trucks and other mobile businesses. Approximately 3,000 independent operators are licensed in the U.S. by the FCC to provide this service. Nextel Communications has acquired a number of SMR operators and turned them into a nationwide system.

Simple Network Management Protocol (SNMP)

A widely used network monitoring and control protocol.

Simple Object Access Protocol (SOAP)

A way for a program running in one kind of operating system (such as Windows 2000) to communicate with a program in the same or another kind of an operating system (such as Linux) by using the World Wide Web's Hypertext Transfer Protocol (HTTP) and its Extensible Markup Language (XML) as the mechanisms for information exchange. Since Web protocols are installed and available for use by all major operating system platforms, HTTP and XML provide an already at-hand solution to the problem of how programs running under different operating systems in a network can communicate with each other. SOAP specifies exactly how to encode an HTTP header and an XML file so that a program in one computer can call a program in another computer and pass it information. It also specifies how the called program can return a response.

Speech Recognition

Ability of machines to respond to spoken commands. Speech recognition enables hands-free control of various electronic devices and the automatic creation of print-ready dictation. Early applications for speech recognition were automated telephone systems and medical dictation software. It is still used for dictation, querying databases, and for giving commands to computer-based systems, especially in professions that rely on specialized vocabularies.

Structured Query Language (SQL)

A standard interactive and programming language for getting information from and updating a database.

Time Division Multiple Access (TDMA)

A satellite and cellular phone technology that weaves multiple digital signals into a single high-speed channel. For cellular, TDMA triples the capacity of the original analog method (FDMA). It divides each channel into three time slots in order to increase the amount of data that can be carried. The GSM cellular system is also based on TDMA, but GSM defines the entire network, not just the air interface.

Time Division Multiplexing (TDM)

Technology that transmits multiple signals simultaneously over a single transmission path. Each lower-speed signal is time sliced into one high-speed transmission. The receiving end divides the single stream back into its original signals. TDM enabled the telephone companies to migrate from analog to digital on all their long distance trunks. The technology is used in channel banks, which convert 24 analog voice conversations into one digital T1 line.

Telemetry

An automated communications process that enables measurements to be made and other data to be collected at remote locations, and transmitted to receiving equipment for monitoring, display, and recording. When telemetry was new; information was sent over wires. Today it uses radio transmission. The process is the same in either case. Major applications include monitoring electric-power plants, gathering meteorological data, and monitoring manned and unmanned space flights.

Teleworking

Teleworking and telecommuting are synonyms for the use of telecommunication to work outside the traditional office or workplace, usually at home or in a mobile situation.

Third Generation (3G) Wireless Networks

Advanced generation of systems, which combines high-speed data transmission with Internet Protocol (IP) based services in fixed, portable, and mobile environments.

3rd Generation Partnership Project (3GPP)

A worldwide consortium of standards organizations (ARIB, CWTS, ETSI, T1, TTA, and TTC) that is developing the technical specifications for IMT-2000. 3GPP develops the W-CDMA technology.

Transmission Control Protocol/Internet Protocol (TCP/IP)

The basic communication language or protocol of the Internet. It can also be used as a communications protocol in a private network (either an intranet or an extranet). TCP/IP is a two-layer program. The higher layer, Transmission Control Protocol, manages the assembling of a message or file into smaller packets that are transmitted over the Internet and received by a TCP layer that reassembles the packets into the original message. The lower layer, Internet Protocol, handles the address part of each packet so that it gets to the right destination. Each gateway computer on the network checks this address to see where to forward the message. Even though some packets from the same message are routed differently than others, they'll be reassembled at the destination.

Universal Product Code (UPC)

The standard bar code printed on retail merchandise. It contains the vendor's identification number and the product number, which is read by passing the bar code over a scanner.

Universal Wireless Communications Consortium (UWCC)

An organization whose members support the IS-41 (WIN) information and control system and EDGE enhancements for increased data rates.

Very Small Aperture Satellite Terminal (VSAT)

A small earth station for satellite transmission that handles up to 56 Kbps of digital transmission. VSATs that handle an Earth-bound T1 data rate (up to 1.544 Mbps) are called "TSATs."

Virtual Private Network (VPN)

A VPN is a private data network that makes use of the public telecommunication infrastructure, maintaining privacy through the use of a tunneling protocol and security procedures. Using a virtual private network involves encrypting data before sending it through the public network and decrypting it at the receiving end. Common carriers have built VPNs for years that appear as private national or international networks to the customer, but that share backbone trunks with other customers.

Voice Over IP (VoIP)

VoIP (that is, voice delivered using the Internet Protocol) is a term used in IP telephony for a set of facilities for managing the delivery of voice information using IP. In general, this means sending voice information in digital form in discrete packets rather than in the traditional circuit-committed protocols of the public switched telephone network. A major advantage of VoIP and Internet telephony is that it avoids the tolls charged by ordinary telephone service.

Wide Area Network (WAN)

A data network typically extending a LAN outside the building, over telephone common carrier lines to link to other LANs in remote buildings, many times geographically to dispersed in other cities or countries. A WAN typically uses common carrier lines (typically leased lines) which LANs do not.

Wideband Code-Division Multiple Access (WCDMA)

An ITU standard derived from code-division multiple access (CDMA) that is officially known as IMT-2000 direct spread. WCDMA is a third-generation (3G) mobile wireless technology offering much higher data speeds to mobile and portable wireless devices than commonly offered in today's market. WCDMA can support mobile/portable voice, images, data, and video communications at up to 2 Mbps (local area access) or 384 Kbps (wide area access). The input signals are digitized and transmitted in coded, spread-spectrum mode over a broad range of frequencies. A 5 MHz-wide carrier is used, compared with 200 KHz-wide carrier for narrowband CDMA.

WINSOCK (WINdows SOCKets)

The Windows interface to TCP/IP, which is the communications protocol of UNIX networks and the Internet. Windows network applications that communicate via TCP/IP are Winsock compliant, as are the implementations of the TCP/IP protocol (TCP/IP stacks) from Microsoft or from third parties.

Wireless Access Protocol (WAP)

A specification for a set of communication protocols to standardize the way that wireless devices (e.g., cellular phones, pagers, PDAs) can be used for Internet access, including e-mail, the World Wide Web, newsgroups, and Internet Relay Chat.

Wireless Application Service Provider (WASP)

A WASP is part of a growing industry sector resulting from the convergence of two trends – wireless communications and the outsourcing of services. A WASP performs the same service for wireless clients as a regular application service provider (ASP) does for wired clients – it provides Web-based access to applications and services that would otherwise have to be stored locally. The main difference with WASP is that it enables customers to access the service from a variety of wireless devices, such as the smartphone and the personal digital assistant.

Although the business world is increasingly mobile, many corporations are resisting the change, because of concerns about set-up and maintenance costs and the need for in-house expertise. WASPs offer businesses the advantages of wireless service with less expense and fewer risks. Because mobile applications are subscribed to, rather than purchased, up-front costs are lower; because the WASP provides support, staffing and training costs are lower.

Wireless Local Area Network (WLAN)

A wireless LAN is one in which a mobile user can connect to a LAN through a wireless (radio) connection. A standard, IEEE 802.11, specifies the technologies for wireless LANs.

Wireless Markup Language (WML)

A tag-based language used in the Wireless Application Protocol (WAP), that allows the text portions of web pages to be presented on cellular phones and personal digital assistants (PDA's) via wireless access.

X.25

The X.25 protocol, adopted as a standard by the Consultative Committee for International Telegraph and Telephone (CCITT), is a commonly used network protocol. The X.25 protocol allows computers on different public networks (such as CompuServe, Tymnet, or a TCP/IP network) to communicate through an intermediary computer at the network layer level. X.25's protocols correspond closely to the data-link and physical-layer protocols defined in the Open Systems Interconnection (OSI) communication model.

eXtensible Markup Language (XML)

A markup language for documents containing structured information. Structured information contains both content (words, pictures, etc.) and some indication of what role that content plays (for example, content in a section heading has a different meaning than content in a footnote, a figure, or a caption, or database table, etc.). Almost all documents have some structure. A markup language is a mechanism to identify structures in a document. The XML specification defines a standard way to add markup to documents.

INDUSTRY DIRECTORY

The following is a partial list of industry vendors. While we recognize that this is not comprehensive because of the ever-changing nature of the industry, it was not our intention to exclude any particular company.

@Hand Corporation
8834 Capital of Texas Highway
Suite 210
Austin, TX 7875
(512) 231-9993
www.hand.com

At Road, Inc.
47200 Bayside Parkway
Freemont, CA 94538
(510) 668-1638
www.atroad.com

2Roam, Inc.
2686 Middlefield Road
Unit 1A
Redwood City, CA 94063
(650) 480-1100
www.2roam.com

Aeris.net
1245 S. Winchester Boulevard
2nd Floor
San Jose, CA 95128
(408) 557-1900
www.aeris.net

Aether Systems, Inc.
11460 Cronridge Drive
Owings Mills, MD 21117
(410) 654-6400
www.aethersystems.com

Air2Web, Inc.
1230 Peachtree Street, NE
Promenade II, Suite 1200
Atlanta, GA 30309
(404) 815-7707
www.air2web.com

Anywhereyougo.com
3008 Taylor Street
Dallas, TX 75226
(214) 752-0084
www.anywhereyougo.com

AT&T Wireless Services, Inc.
16221 NE 72nd Way
Redmond, WA 98052
(425) 580-6000
www.attws.com

Avant Go.
25881, Industrial Blvd.
Hayward, CA 94545
(510) 259-4000
www.avantgo.com

Brience, Inc.
128 Spear Street
3rd Floor
San Francisco, CA 94105
(415) 974-5300
www.brience.com

Broadbeam Corporation
100 College Road West
Princeton, NJ 08540
(609) 734-0300
www.broadbeam.com

Cellular Telecommunications & Internet Association (CTIA)
1250 Connecticut Ave, NW
Suite 800
Washington, DC 20036
(202) 785-0081
www.wow-com.com

Certicom Corp.
5520 Explorer Drive
4th Floor
Mississauga, L4W 5L1
Canada
(905) 507-4220
www.certicom.com

Cingular Interactive, LP
10 Woodbridge Center Drive
Woodbridge, NJ 07095
(888) 825-2727
www.bellsouthwd.com

Compaq Computer Corporation
20555 SH 249
Houston, Texas 77070
281-370-0670
www.compaq.com

Diversinet Corporation
2225 Sheppard Avenue East
Suite 1700
Toronto, ON M2J 5C2
Canada
(416) 756-2324
www.dvnet.com

Ericsson Cyberlab NY
55 Broad Street
16th Floor
New York, NY 10004
(212) 612-1299
www.ericsson.com/cyberlab

Everypath, Inc.
2211 North First Street
Suite 200
San Jose, CA 95131
(408) 562-8000
www.everypath.com

Extended Systems
5777 N. Meeker Avenue
Boise, ID 83713
(208) 322-7575
www.extendedsystems.com

Fujitsu PC Corporation
5200 Patrick Henry Drive
Santa Clara, CA 95054
(408) 982-9500
www.fujitsupc.com

Geoworks Corporation
960 Atlanta Avenue
Alameda, CA 94501
(510) 814-5811
www.geoworks.com

Global Wireless Data, LLC
3000-B Business Park Drive
Norcross, GA 30071
(770) 447-4990
www.globalwirelessdata.com

GoAmerica Communications Corp.
401 Hackensack Avenue
Hackensack, NJ 07601
(201) 966-1717
www.goamerica.net

Handango.com, Inc.
305 NE Loop 820
Suite 200
Hurst, TX 76053
(817) 280-0129
www.handago.com

Hewlett-Packard
3000 Hanover Street
Palo Alto, CA 94304
(650) 857-1501
www.hp.com

i3 Mobile, Inc.
181 Harbor Drive
Stamford, CT 06902
(203) 428-3000
www.i3mobile.com

IBM Corporation
New Orchard Road
Armonk, NY 10504
(914) 499-1900
www.ibm.com

iConverse
71 Second Avenue
Third Floor
Waltham, MA 02451
(781) 622-5050
www.iconverse.com

Imedeon, Inc.
11605 Haynes Bridge Road
Suite 600
Alpharetta, GA 30004
(770) 777-8100
www.imedeon.com

Infowave Software, Inc.
4664 Lougheed Highway
Suite 200
Burnaby, BC V5C 5T5
Canada
(604) 473-3600
www.infowave.com

Intermec Technologies Corporation
6001 36th Avenue West
P.O Box 4280
Everett, WA 98203
(425) 348-2600
www.intermec.com

Itronix Corporation
South 801 Stevens
Spokane, WA 99204
Phone: 509-624-6600
www.itronix.com

JP Mobile, Inc.
12000 Ford Road
Suite 400
Dallas, TX 75234
(972) 484-5432
www.jpmobile.com

Metricom, Inc.
333 West Julian Street
San Jose, CA 95110
(408) 282-3000
www.metricom.com

Microsoft Corporation
One Microsoft Way
Redmond, WA 98052
(425) 882-8080
www.microsoft.com

MindMatrix, Inc.
5001 Baum Boulevard
Pittsburgh, PA 15213
(412) 683-0222
www.mindmatrix.net

Mobile Insights, Inc.
2001 Landings Drive
Mountain View, CA 94043
(650) 390-9800
www.mobileinsights.com

Motient Corporation
10802 Parkridge Blvd
Reston, VA 20191
(703) 758-6000
www.motient.com

Motorola, Inc.
1303 E. Algonquin Road
Schaumburg, IL 60196
(847) 576-8000
www.motorola.com

Nortel Networks Corporation
2221 Lakeside Boulevard
Richardson, TX 75082
(972) 684-1000
www.nortelnetworks.com

Openwave Systems, Inc.
11201 SE 8th Street
Suite 110
Bellevue, WA 98004
(425) 372-2200
www.software.com

OracleMobile, Inc.
1000 Bridge Parkway
Redwood Shores, CA 94065
(650) 506-7000
www.oraclemobile.com

Palm, Inc. Corporate Headquarters
5470 Great America Pkwy
Santa Clara, CA 95054
(408) 878-9000
www.palm.com

Panasonic/Matsushita Electric Corporation of America
One Panasonic Way
Secaucus, NJ 07094
www.panasonic.com

Personal Communications Industry Association (PCIA)
500 Montgomery Street
Suite 700
Alexandria, VA 22314
Tel: (703) 739-0300
www.pcia.com

Portable Computing and Communication Association (PCCA)
P.O. Box 2460
Boulder Creek, CA 95006
www.pcca.org

Psion PLC
12 Park Crescent
London
W1B 1PH
United Kingdom
+44 (0) 870-608-0680
www.psion.com

Qualcomm, Inc.
5775 Morehouse Drive
San Diego, CA 92121
(858) 587-1121
www.qualcomm.com

Research In Motion Limited
295 Phillip Street
Waterloo, ON N2L 3W8
Canada
(519) 888-7465
www.rim.net

Sierra Wireless, Inc.
13575 Commerce Parkway
Suite 150
Richmond, BC V6V 3A4
Canada
(604) 231-1100
www.sierrawireless.com

Sprint PCS
P.O. BOX 8077
London, KY 40742
www.sprintpcs.com

Symbian Ltd.
Sentinel House
16 Harcourt Street
London W1H 1AD
United Kingdom
+44 (0) 20-7563-2000
www.symbian.com

Symbol Technologies, Inc.
One Symbol Plaza
Holtsville, New York 11742-1300
(631) 738-2400
www.symbol.com

Synchrologic
200 North Point Center East
Suite 600
Alpharetta, GA 30022
(770) 754-5600
info@synchrologic.com
www.synchrologic.com

Verizon Wireless Services, LLC
180 Washington Valley Road
Bedminster, NJ 07921
(908) 306-7000
www.verizonwireless.com

Wireless Developer Network
GeoComm International Corporation
4565 Commercial Drive
Suite "D"
Niceville, FL 32578
(850) 897-0858
www.wirelessdevnet.com

Wireless Matrix Corporation
Sunrise Technology Park
12369-B Sunrise Valley Drive
Reston, VA 20191
(703) 262-0500
www.norcomnetworks.com
Wireless Now
1130 Connecticut Ave., NW
Suite 210
Washington, DC 20036
(202) 530-7600
www.wirelessnow.com

W-Technologies, Inc.
150 Broadway
15th Floor
New York, NY 10038
(212) 406-7685
www.w-technologies.com

Wysdom, Inc.
30 West Beaver Creek Road
Unit 111, 2nd Floor
Toronto, ON L4B 3K1
Canada
(905) 763-6979
www.wysdom.com

Zsigo Wireless, Inc.
2875 Northwind Drive
Suite 232
East Lansing, MI 48823
(517) 337-3995
www.zsigo.com

BIBLIOGRAPHY

Christensen, Clayton M. 1997. *The Innovator's Dilemma* Boston: Harvard Business School Press.

Coster, Paul, 2000. "Out of Thin Air: Emerging Wireless Infrastructure, Software and Services." *Equity Research Report, J.P. Morgan Securities, Inc.*

Emmett, Arielle, 1999. "The Focus is Middleware." *Wireless Integration.* September-October: 22.

Hamel, Gary, 1997. *Leading the Revolution* Boston: Harvard Business School Press.

Lewis, Brenda. "Location, Location, Location: The Killer App for Mobile E-Commerce?" *WirelessAgenda 2000, Session #B6, Conference Proceedings* Wireless Data Forum/Cellular Telecommunications & Internet Association.

Pearce, Alan, 1999. "Is Satellite Telephony Worth Saving".*Wireless Integration* September - October: 19.

Polizzi, Tom, 1999. "The New Meaning of the Word Wireless". *Wireless Integration Magazine Special Report.* July-August.

Stoffels, Robert, 1998. "The CDMA Film Noir" *America's Network* June: 22.

Sweeney, Daniel, 1999. Wireless Local Area Networks (WLANs).*1999 Wireless Products & Services Buyer's Guide, Wireless Integration.* March - April.

Sweeney, Daniel, 1999. "Wireless WANs." *1999 Wireless Products & Services Buyer's Guide, Wireless Integration.* 35-36

Wireless Data Forum/Cellular Telecommunications & Internet Association. 2001. *Going Mobile II: The Wireless Web and other Data Solutions.* Washington, D.C.

INDEX

A

B

E

G

N

R

S